Accuplacer

Subject Test Mathematics

Student Practice Workbook

+ Two Full-Length Accuplacer Math Tests

Math Notion

www.MathNotion.com

Accuplacer Subject Test Mathematics

Published in the United State of America By

The Math Notion

Web: WWW.MathNotion.com

Email: info@Mathnotion.com

ISBN: 978-1-63620-045-3

The Math Notion

Michael Smith has been a math instructor for over a decade now. He launched the Math Notion. Since 2006, we have devoted our time to both teaching and developing exceptional math learning materials. As a test prep company, we have worked with thousands of students. We have used the feedback of our students to develop a unique study program that can be used by students to drastically improve their math scores fast and effectively. We have more than a thousand Math learning books including:

– **SAT Math Prep**

– **ACT Math Prep**

– **GRE Math Prep**

– **PSAT Math Prep**

– **Common Core Math Prep**

–**many Math Education Workbooks, Study Guides, Practice and Exercise Books**

As an experienced Math test preparation company, we have helped many students raise their standardized test scores—and attend the colleges of their dreams: We tutor online and in person, we teach students in large groups, and we provide training materials and textbooks through our website and through Amazon.

You can contact us via email at:

info@Mathnotion.com

Get the Targeted Practice You Need to Ace the Accuplacer Math Test!

Accuplacer Subject Test - Mathematics includes easy-to-follow instructions, helpful examples, and plenty of math practice problems to assist students to master each concept, brush up their problem-solving skills, and create confidence.

The Accuplacer math practice book provides numerous opportunities to evaluate basic skills along with abundant remediation and intervention activities. It is a skill that permits you to quickly master intricate information and produce better leads in less time.

Students can boost their test-taking skills by taking the book's two practice Accuplacer Math exams. All test questions answered and explained in detail.

Important Features of the Accuplacer Math Book:

- A **complete review** of Accuplacer math test topics,
- Over 2,500 practice problems covering all topics tested,
- The most important concepts you need to know,
- Clear and concise, easy-to-follow sections,
- Well designed for enhanced learning and interest,
- Hands-on experience with all question types
- **2 full-length practice tests** with detailed answer explanations
- Cost-Effective Pricing

Powerful math exercises to help you avoid traps and pacing yourself to beat the Accuplacer test. Students will gain valuable experience and raise their confidence by taking math practice tests, learning about test structure, and gaining a deeper understanding of what is tested on the Accuplacer Math. If ever there was a book to respond to the pressure to increase students' test scores, this is it.

WWW.MathNotion.COM

… So Much More Online!

✓ FREE Math Lessons

✓ More Math Learning Books!

✓ Mathematics Worksheets

✓ Online Math Tutors

For a PDF Version of This Book

Please Visit WWW.MathNotion.com

Contents

Chapter 1 :

Integers and Number Theory

Topics that you'll practice in this chapter:

- ✓ Rounding
- ✓ Whole Number Addition and Subtraction
- ✓ Whole Number Multiplication and Division
- ✓ Rounding and Estimates
- ✓ Adding and Subtracting Integers
- ✓ Multiplying and Dividing Integers
- ✓ Order of Operations
- ✓ Ordering Integers and Numbers
- ✓ Integers and Absolute Value
- ✓ Factoring Numbers
- ✓ Greatest Common Factor (GCF)
- ✓ Least Common Multiple (LCM)

"Wherever there is number, there is beauty." –Proclus

Rounding

✍ **Round each number to the nearest ten.**

1) 42 = ____   5) 19 = ____   9) 48 = ____

2) 88 = ____   6) 25 = ____   10) 81 = ____

3) 24 = ____   7) 93 = ____   11) 58 = ____

4) 57 = ____   8) 71 = ____   12) 87 = ____

✍ **Round each number to the nearest hundred.**

13) 198 = ____   17) 321 = ____   21) 580 = ____

14) 387 = ____   18) 433 = ____   22) 868 = ____

15) 816 = ____   19) 579 = ____   23) 480 = ____

16) 101 = ____   20) 825 = ____   24) 287 = ____

✍ **Round each number to the nearest thousand.**

25) 1,382 = ____   29) 9,099 = ____   33) 52,866 = ____

26) 3,420 = ____   30) 22,980 = ____   34) 85,190 = ____

27) 4,254 = ____   31) 45,188 = ____   35) 70,990 = ____

28) 6,861 = ____   32) 16,808 = ____   36) 26,869 = ____

Rounding and Estimates

🖎 **Estimate the sum by rounding each number to the nearest ten.**

1) $13 + 22 =$ _____

2) $71 + 23 =$ _____

3) $61 + 58 =$ _____

4) $56 + 85 =$ _____

5) $368 + 249 =$ _____

6) $330 + 903 =$ _____

7) $471 + 293 =$ _____

8) $1,950 + 2,655 =$ _____

🖎 **Estimate the product by rounding each number to the nearest ten.**

9) $32 \times 71 =$ _____

10) $12 \times 33 =$ _____

11) $31 \times 83 =$ _____

12) $19 \times 11 =$ _____

13) $42 \times 76 =$ _____

14) $63 \times 34 =$ _____

15) $19 \times 31 =$ _____

16) $59 \times 71 =$ _____

🖎 **Estimate the sum or product by rounding each number to the nearest ten.**

17) $\begin{array}{r} 29 \\ \times\,12 \\ \hline \end{array}$

19) $\begin{array}{r} 48 \\ +\,82 \\ \hline \end{array}$

21) $\begin{array}{r} 37 \\ \times\,14 \\ \hline \end{array}$

18) $\begin{array}{r} 37 \\ \times\,26 \\ \hline \end{array}$

20) $\begin{array}{r} 65 \\ +\,44 \\ \hline \end{array}$

22) $\begin{array}{r} 71 \\ +\,32 \\ \hline \end{array}$

Adding and Subtracting Integers

✎ **Find each sum.**

1) $14 + (-6) =$

2) $(-13) + (-20) =$

3) $5 + (-28) =$

4) $50 + (-12) =$

5) $(-7) + (-15) + 3 =$

6) $30 + (-14) + 8 =$

7) $40 + (-10) + (-14) + 17 =$

8) $(-15) + (-20) + 13 + 35 =$

9) $40 + (-20) + (38 - 29) =$

10) $28 + (-12) + (30 - 12) =$

✎ **Find each difference.**

11) $(-18) - (-7) =$

12) $25 - (-14) =$

13) $(-20) - 36 =$

14) $34 - (-19) =$

15) $51 - (30 - 21) =$

16) $17 - (5) - (-24) =$

17) $(35 + 20) - (-46) =$

18) $48 - 16 - (-8) =$

19) $62 - (28 + 17) - (-15) =$

20) $58 - (-23) - (-31) =$

21) $19 - (-8) - (-13) =$

22) $(19 - 24) - (-14) =$

23) $27 - 33 - (-21) =$

24) $58 - (32 + 24) - (-9) =$

25) $36 - (-30) + (-17) =$

26) $27 - (-42) + (-31) =$

Multiplying and Dividing Integers

✎ **Find each product.**

1) $(-9) \times (-5) =$

2) $(-3) \times 9 =$

3) $8 \times (-12) =$

4) $(-7) \times (-20) =$

5) $(-3) \times (-5) \times 6 =$

6) $(14 - 3) \times (-8) =$

7) $12 \times (-9) \times (-3) =$

8) $(140 + 10) \times (-2) =$

9) $10 \times (-12 + 8) \times 3 =$

10) $(-8) \times (-5) \times (-10) =$

✎ **Find each quotient.**

11) $42 \div (-7) =$

12) $(-48) \div (-6) =$

13) $(-40) \div (-8) =$

14) $54 \div (-2) =$

15) $152 \div 19 =$

16) $(-144) \div (-12) =$

17) $180 \div (-10) =$

18) $(-312) \div (-12) =$

19) $221 \div (-13) =$

20) $(-126) \div (6) =$

21) $(-161) \div (-7) =$

22) $-266 \div (-14) =$

23) $(-120) \div (-4) =$

24) $270 \div (-18) =$

25) $(-208) \div (-8) =$

26) $(135) \div (-15) =$

Order of Operations

✎ Evaluate each expression.

1) $7 + (5 \times 4) =$

2) $14 - (3 \times 6) =$

3) $(19 \times 4) + 16 =$

4) $(16 - 7) - (8 \times 2) =$

5) $27 + (18 \div 3) =$

6) $(18 \times 8) \div 6 =$

7) $(32 \div 4) \times (-2) =$

8) $(9 \times 4) + (32 - 18) =$

9) $24 + (4 \times 3) + 7 =$

10) $(36 \times 3) \div (2 + 2) =$

11) $(-7) + (12 \times 3) + 11 =$

12) $(8 \times 5) - (24 \div 6) =$

13) $(7 \times 6 \div 3) - (12 + 9) =$

14) $(13 + 5 - 14) \times 3 - 2 =$

15) $(20 - 14 + 30) \times (64 \div 4) =$

16) $32 + \big(28 - (36 \div 9)\big) =$

17) $(7 + 6 - 4 - 7) + (15 \div 5) =$

18) $(85 - 20) + (20 - 18 + 7) =$

19) $(20 \times 2) + (14 \times 3) - 22 =$

20) $18 + 5 - (30 \times 3) + 20 =$

Ordering Integers and Numbers

✎ **Order each set of integers from least to greatest.**

1) $8, -10, -5, -3, 4$ ___, ___, ___, ___, ___, ___

2) $-10, -18, 6, 14, 27$ ___, ___, ___, ___, ___, ___

3) $15, -8, -21, 21, -23$ ___, ___, ___, ___, ___, ___

4) $-14, -40, 23, -12, 47$ ___, ___, ___, ___, ___, ___

5) $59, -54, 32, -57, 36$ ___, ___, ___, ___, ___, ___

6) $68, 26, -19, 47, -34$ ___, ___, ___, ___, ___, ___

✎ **Order each set of integers from greatest to least.**

7) $18, 36, -16, -18, -10$ ___, ___, ___, ___, ___, ___

8) $27, 34, -12, -24, 94$ ___, ___, ___, ___, ___, ___

9) $50, -21, -13, 42, -2$ ___, ___, ___, ___, ___, ___

10) $37, 46, -20, -16, 86$ ___, ___, ___, ___, ___, ___

11) $-18, 88, -26, -59, 75$ ___, ___, ___, ___, ___, ___

12) $-65, -30, -25, 3, 14$ ___, ___, ___, ___, ___, ___

Integers and Absolute Value

✎ **Write absolute value of each number.**

1) $|-2| =$

2) $|-27| =$

3) $|-20| =$

4) $|14| =$

5) $|6| =$

6) $|-55| =$

7) $|16| =$

8) $|2| =$

9) $|54| =$

10) $|-4| =$

11) $|-11|$

12) $|88| =$

13) $|0| =$

14) $|79| =$

15) $|-32| =$

16) $|-17| =$

17) $|42| =$

18) $|-46| =$

19) $|1| =$

20) $|-40| =$

✎ **Evaluate the value.**

21) $|-5| - \dfrac{|-21|}{7} =$

22) $14 - |3 - 15| - |-4| =$

23) $\dfrac{|-32|}{4} \times |-4| =$

24) $\dfrac{|7 \times (-3)|}{7} \times \dfrac{|-19|}{3} =$

25) $|4 \times (-5)| + \dfrac{|-40|}{5} =$

26) $\dfrac{|-45|}{9} \times \dfrac{|-24|}{12} =$

27) $|-12 + 8| \times \dfrac{|-7 \times 7|}{7} =$

28) $\dfrac{|-11 \times 2|}{4} \times |-16| =$

Factoring Numbers

✎ **List all positive factors of each number.**

1) 9

2) 16

3) 24

4) 30

5) 26

6) 46

7) 20

8) 68

9) 28

10) 98

11) 14

12) 54

13) 55

14) 18

15) 63

16) 34

17) 50

18) 62

19) 95

20) 64

21) 70

22) 45

23) 22

24) 65

Greatest Common Factor

✍ **Find the GCF for each number pair.**

1) 6, 2	9) 12, 18	17) 42, 14
2) 4, 5	10) 4, 36	18) 16, 40
3) 3, 12	11) 6, 10	19) 9, 2, 3
4) 7, 3	12) 28, 52	20) 5, 15, 10
5) 5, 10	13) 25, 10	21) 7, 9, 2
6) 8, 48	14) 22, 24	22) 16, 64
7) 6, 18	15) 9, 54	23) 30, 48
8) 9, 15	16) 8, 54	24) 36, 63

Least Common Multiple

✎ Find the LCM for each number pair.

1) 6, 9

2) 15, 45

3) 16, 40

4) 12, 36

5) 18, 27

6) 14, 42

7) 6, 30

8) 8, 56

9) 7, 21

10) 8, 20

11) 15, 25

12) 7, 9

13) 4, 11

14) 8, 28

15) 28, 56

16) 40, 50

17) 12, 13

18) 22, 11

19) 36, 20

20) 15, 35

21) 18, 81

22) 30, 54

23) 18, 45

24) 75, 25

Answers of Worksheets

Rounding

1) 40	10) 80	19) 600	28) 7,000
2) 90	11) 60	20) 800	29) 9,000
3) 20	12) 90	21) 600	30) 23,000
4) 60	13) 200	22) 900	31) 45,000
5) 20	14) 400	23) 500	32) 17,000
6) 30	15) 800	24) 300	33) 53,000
7) 90	16) 100	25) 1,000	34) 85,000
8) 70	17) 300	26) 3,000	35) 71,000
9) 50	18) 400	27) 4,000	36) 27,000

Rounding and Estimates

1) 30	7) 760	13) 3,200	19) 130
2) 90	8) 4,610	14) 1,800	20) 110
3) 120	9) 2,100	15) 600	21) 400
4) 150	10) 300	16) 4,200	22) 100
5) 620	11) 2,400	17) 300	
6) 1,230	12) 200	18) 1,200	

Adding and Subtracting Integers

1) 8	8) 13	15) 42	22) 9
2) −33	9) 29	16) 36	23) 15
3) −23	10) 34	17) 101	24) 11
4) 38	11) −11	18) 40	25) 49
5) −19	12) 39	19) 32	26) 38
6) 24	13) −56	20) 112	
7) 33	14) 53	21) 40	

Multiplying and Dividing Integers

1) 45	6) −88	11) −6	16) 12
2) −27	7) 324	12) 8	17) −18
3) −96	8) −300	13) 5	18) 26
4) 140	9) −120	14) −27	19) −17
5) 90	10) −400	15) 8	20) −21

21) 23
22) 19
23) 30
24) −15
25) 26
26) −9

Order of Operations

1) 27
2) −4
3) 92
4) −7
5) 33
6) 24
7) −16
8) 50
9) 43
10) 27
11) 40
12) 36
13) −7
14) 10
15) 576
16) 56
17) 5
18) 74
19) 60
20) −47

Ordering Integers and Numbers

1) $-10, -5, -3, 4, 8$
2) $-18, -10, 6, 14, 27$
3) $-23, -21, -8, 15, 21$
4) $-40, -14, -12, 23, 47$
5) $-57, -54, 32, 36, 59$
6) $-34, -19, 26, 47, 68$
7) $36, 18, -10, -16, -18$
8) $94, 34, 27, -12, -24$
9) $50, 42, -2, -13, -21$
10) $86, 46, 37, -16, -20$
11) $88, 75, -18, -26, -59$
12) $14, 3, -25, -30, -65$

Integers and Absolute Value

1) 2
2) 27
3) 20
4) 14
5) 6
6) 55
7) 16
8) 2
9) 54
10) 4
11) 11
12) 88
13) 0
14) 79
15) 32
16) 17
17) 42
18) 46
19) 1
20) 40
21) 2
22) −2
23) 32
24) 19
25) 28
26) 10
27) 28
28) 88

Factoring Numbers

1) $1, 3, 9$
2) $1, 2, 4, 8, 16$
3) $1, 2, 3, 4, 6, 8, 12, 24$
4) $1, 2, 3, 5, 6, 10, 15, 30$
5) $1, 2, 13, 26$
6) $1, 2, 23, 46$
7) $1, 2, 4, 5, 10, 20$
8) $1, 2, 4, 17, 34, 68$
9) $1, 2, 4, 7, 14, 28$
10) $1, 2, 7, 14, 49, 98$
11) $1, 2, 7, 14$
12) $1, 2, 3, 6, 9, 18, 27, 54$
13) $1, 5, 11, 55$
14) $1, 2, 3, 6, 9, 18$
15) $1, 3, 7, 9, 21, 63$
16) $1, 2, 17, 34$
17) $1, 2, 5, 10, 25, 50$
18) $1, 2, 31, 62$
19) $1, 5, 19, 95$
20) $1, 2, 4, 8, 16, 32, 64$
21) $1, 2, 5, 7, 10, 14, 35, 70$
22) $1, 3, 5, 9, 15, 45$
23) $1, 2, 11, 22$
24) $1, 5, 13, 65$

Greatest Common Factor

1) 2	7) 6	13) 5	19) 1
2) 1	8) 3	14) 2	20) 5
3) 3	9) 6	15) 9	21) 1
4) 1	10) 4	16) 2	22) 16
5) 5	11) 2	17) 14	23) 6
6) 8	12) 4	18) 8	24) 9

Least Common Multiple

1) 18	7) 30	13) 44	19) 180
2) 45	8) 56	14) 56	20) 105
3) 80	9) 21	15) 56	21) 162
4) 36	10) 40	16) 200	22) 270
5) 54	11) 75	17) 156	23) 90
6) 42	12) 63	18) 22	24) 75

Chapter 2 :

Fractions and Decimals

Topics that you'll practice in this chapter:

- ✓ Simplifying Fractions
- ✓ Adding and Subtracting Fractions
- ✓ Multiplying and Dividing Fractions
- ✓ Adding and Subtract Mixed Numbers
- ✓ Multiplying and Dividing Mixed Numbers
- ✓ Adding and Subtracting Decimals
- ✓ Multiplying and Dividing Decimals
- ✓ Comparing Decimals
- ✓ Rounding Decimals

"A Man is like a fraction whose numerator is what he is and whose denominator is what he thinks of himself. The larger the denominator, the smaller the fraction." –Tolstoy

Simplifying Fractions

✎ **Simplify each fraction to its lowest terms.**

1) $\dfrac{5}{10} =$

2) $\dfrac{28}{35} =$

3) $\dfrac{27}{36} =$

4) $\dfrac{40}{80} =$

5) $\dfrac{14}{56} =$

6) $\dfrac{32}{48} =$

7) $\dfrac{52}{65} =$

8) $\dfrac{15}{60} =$

9) $\dfrac{80}{160} =$

10) $\dfrac{55}{77} =$

11) $\dfrac{28}{112} =$

12) $\dfrac{32}{64} =$

13) $\dfrac{63}{72} =$

14) $\dfrac{81}{90} =$

15) $\dfrac{35}{105} =$

16) $\dfrac{25}{70} =$

17) $\dfrac{80}{280} =$

18) $\dfrac{12}{81} =$

19) $\dfrac{36}{186} =$

20) $\dfrac{240}{540} =$

21) $\dfrac{70}{560} =$

✎ **Find the answer for each problem.**

22) Which of the following fractions equal to $\dfrac{3}{4}$? ____

    A. $\dfrac{60}{90}$
    B. $\dfrac{43}{104}$
    C. $\dfrac{48}{64}$
    D. $\dfrac{150}{300}$

23) Which of the following fractions equal to $\dfrac{5}{8}$? ____

    A. $\dfrac{125}{200}$
    B. $\dfrac{115}{200}$
    C. $\dfrac{50}{100}$
    D. $\dfrac{30}{90}$

24) Which of the following fractions equal to $\dfrac{3}{7}$? ____

    A. $\dfrac{58}{116}$
    B. $\dfrac{54}{126}$
    C. $\dfrac{270}{167}$
    D. $\dfrac{42}{63}$

Adding and Subtracting Fractions

✎**Find the sum.**

1) $\frac{5}{9} + \frac{4}{9} =$

5) $\frac{1}{4} + \frac{3}{5} =$

9) $\frac{5}{7} + \frac{2}{3} =$

2) $\frac{1}{2} + \frac{1}{7} =$

6) $\frac{7}{8} + \frac{3}{8} =$

10) $\frac{7}{12} + \frac{3}{4} =$

3) $\frac{3}{8} + \frac{1}{4} =$

7) $\frac{1}{2} + \frac{7}{10} =$

11) $\frac{5}{6} + \frac{2}{5} =$

4) $\frac{3}{5} + \frac{1}{2} =$

8) $\frac{2}{5} + \frac{2}{3} =$

12) $\frac{1}{12} + \frac{2}{3} =$

✎**Find the difference.**

13) $\frac{1}{3} - \frac{1}{6} =$

19) $\frac{5}{6} - \frac{1}{9} =$

25) $\frac{6}{7} - \frac{3}{4} =$

14) $\frac{3}{4} - \frac{1}{8} =$

20) $\frac{3}{4} - \frac{1}{6} =$

26) $\frac{4}{5} - \frac{1}{8} =$

15) $\frac{1}{2} - \frac{1}{3} =$

21) $\frac{7}{8} - \frac{1}{12} =$

27) $\frac{4}{7} - \frac{2}{35} =$

16) $\frac{1}{4} - \frac{1}{5} =$

22) $\frac{8}{15} - \frac{3}{5} =$

28) $\frac{9}{16} - \frac{2}{8} =$

17) $\frac{5}{8} - \frac{2}{3} =$

23) $\frac{3}{12} - \frac{1}{14} =$

29) $\frac{8}{9} - \frac{7}{18} =$

18) $\frac{1}{4} - \frac{1}{7} =$

24) $\frac{10}{13} - \frac{7}{26} =$

30) $\frac{1}{2} - \frac{4}{9} =$

Multiplying and Dividing Fractions

✎ Find the value of each expression in lowest terms.

1) $\frac{1}{5} \times \frac{15}{5} =$

5) $\frac{1}{5} \times \frac{1}{4} =$

9) $\frac{5}{8} \times \frac{3}{5} =$

2) $\frac{9}{12} \times \frac{4}{9} =$

6) $\frac{7}{9} \times \frac{1}{7} =$

10) $\frac{4}{7} \times \frac{1}{8} =$

3) $\frac{1}{16} \times \frac{8}{10} =$

7) $\frac{6}{7} \times \frac{1}{3} =$

11) $\frac{7}{15} \times \frac{5}{7} =$

4) $\frac{1}{24} \times \frac{8}{10} =$

8) $\frac{2}{8} \times \frac{2}{8} =$

12) $\frac{3}{10} \times \frac{5}{9} =$

✎ Find the value of each expression in lowest terms.

13) $\frac{1}{4} \div \frac{1}{8} =$

19) $\frac{2}{7} \div \frac{7}{13} =$

25) $\frac{1}{9} \div \frac{2}{5} =$

14) $\frac{1}{10} \div \frac{1}{5} =$

20) $\frac{1}{24} \div \frac{3}{16} =$

26) $\frac{5}{12} \div \frac{3}{5} =$

15) $\frac{3}{4} \div \frac{1}{5} =$

21) $\frac{7}{12} \div \frac{5}{6} =$

27) $\frac{3}{20} \div \frac{1}{6} =$

16) $\frac{1}{3} \div \frac{5}{6} =$

22) $\frac{22}{18} \div \frac{11}{9} =$

28) $\frac{8}{20} \div \frac{3}{4} =$

17) $\frac{1}{7} \div \frac{8}{42} =$

23) $\frac{9}{35} \div \frac{3}{7} =$

29) $\frac{5}{6} \div \frac{2}{9} =$

18) $\frac{3}{4} \div \frac{1}{6} =$

24) $\frac{2}{7} \div \frac{8}{21} =$

30) $\frac{5}{11} \div \frac{3}{4} =$

Adding and Subtracting Mixed Numbers

✎ **Find the sum.**

1) $3\frac{1}{3} + 2\frac{1}{6} =$

2) $4\frac{1}{2} + 3\frac{1}{2} =$

3) $3\frac{3}{8} + 1\frac{1}{8} =$

4) $2\frac{1}{4} + 2\frac{1}{3} =$

5) $3\frac{5}{6} + 2\frac{7}{12} =$

6) $5\frac{4}{15} + 3\frac{3}{5} =$

7) $2\frac{1}{3} + 4\frac{3}{7} =$

8) $3\frac{1}{2} + 4\frac{2}{5} =$

9) $5\frac{2}{5} + 6\frac{3}{7} =$

10) $8\frac{5}{16} + 6\frac{1}{12} =$

✎ **Find the difference.**

11) $3\frac{1}{4} - 1\frac{3}{4} =$

12) $6\frac{3}{5} - 4\frac{2}{5} =$

13) $4\frac{1}{3} - 3\frac{1}{9} =$

14) $7\frac{1}{7} - 5\frac{1}{2} =$

15) $5\frac{1}{3} - 2\frac{1}{12} =$

16) $8\frac{1}{5} - 4\frac{1}{3} =$

17) $9\frac{1}{4} - 6\frac{1}{8} =$

18) $11\frac{7}{15} - 8\frac{3}{5} =$

19) $14\frac{5}{6} - 11\frac{3}{5} =$

20) $18\frac{2}{7} - 14\frac{1}{5} =$

21) $9\frac{1}{3} - 4\frac{1}{4} =$

22) $6\frac{1}{8} - 4\frac{1}{16} =$

23) $19\frac{3}{8} - 15\frac{1}{3} =$

24) $11\frac{1}{9} - 8\frac{1}{8} =$

25) $17\frac{1}{7} - 11\frac{1}{5} =$

26) $16\frac{2}{9} - 9\frac{5}{7} =$

Multiplying and Dividing Mixed Numbers

✍ **Find the product.**

1) $5\frac{1}{2} \times 2\frac{1}{4} =$

2) $5\frac{1}{3} \times 4\frac{1}{3} =$

3) $5\frac{3}{4} \times 6\frac{1}{4} =$

4) $3\frac{1}{3} \times 2\frac{3}{5} =$

5) $4\frac{8}{10} \times 1\frac{1}{24} =$

6) $6\frac{2}{7} \times 1\frac{1}{11} =$

7) $8\frac{2}{3} \times 3\frac{1}{2} =$

8) $3\frac{4}{7} \times 2\frac{1}{5} =$

9) $5\frac{2}{8} \times 4\frac{1}{6} =$

10) $7\frac{3}{3} \times 1\frac{3}{8} =$

✍ **Find the quotient.**

11) $2\frac{2}{5} \div 4\frac{1}{5} =$

12) $4\frac{1}{6} \div 3\frac{1}{3} =$

13) $6\frac{1}{3} \div 1\frac{1}{2} =$

14) $7\frac{1}{10} \div 2\frac{2}{5} =$

15) $3\frac{1}{3} \div 1\frac{1}{9} =$

16) $1\frac{1}{10} \div 4\frac{1}{2} =$

17) $1\frac{3}{16} \div 5\frac{1}{4} =$

18) $4\frac{1}{3} \div 4\frac{3}{4} =$

19) $9\frac{1}{3} \div 2\frac{1}{4} =$

20) $15\frac{1}{3} \div 5\frac{1}{2} =$

21) $4\frac{1}{6} \div 1\frac{1}{5} =$

22) $1\frac{1}{18} \div 1\frac{2}{9} =$

23) $4\frac{2}{7} \div 1\frac{3}{10} =$

24) $7\frac{1}{3} \div 2\frac{2}{11} =$

25) $8\frac{2}{5} \div 1\frac{1}{6} =$

26) $9\frac{1}{3} \div 2\frac{1}{7} =$

Adding and Subtracting Decimals

✎ **Add and subtract decimals.**

$$
1)\quad \begin{array}{r} 35.19 \\ -\ 24.28 \\ \hline \end{array}
\qquad
4)\quad \begin{array}{r} 38.72 \\ -\ 21.68 \\ \hline \end{array}
\qquad
7)\quad \begin{array}{r} 86.09 \\ -\ 35.14 \\ \hline \end{array}
$$

$$
2)\quad \begin{array}{r} 34.29 \\ +\ 42.58 \\ \hline \end{array}
\qquad
5)\quad \begin{array}{r} 57.39 \\ +\ 26.54 \\ \hline \end{array}
\qquad
8)\quad \begin{array}{r} 54.51 \\ +\ 32.66 \\ \hline \end{array}
$$

$$
3)\quad \begin{array}{r} 61.20 \\ +\ 33.75 \\ \hline \end{array}
\qquad
6)\quad \begin{array}{r} 70.24 \\ -\ 42.35 \\ \hline \end{array}
\qquad
9)\quad \begin{array}{r} 114.21 \\ -\ 88.69 \\ \hline \end{array}
$$

✎ **Find the missing number.**

10) $\_\_\_ + 2.8 = 5.4$

11) $4.1 + \_\_\_ = 5.88$

12) $6.45 + \_\_\_ = 8$

13) $7.25 - \_\_\_ = 3.40$

14) $\_\_\_ - 2.35 = 4.25$

15) $\_\_\_ - 19.85 = 6.54$

16) $22.15 + \_\_\_ = 28.95$

17) $\_\_\_ - 37.16 = 9.42$

18) $\_\_\_ + 24.50 = 34.19$

19) $72.40 + \_\_\_ = 125.20$

Multiplying and Dividing Decimals

✏ **Find the product.**

1) $0.5 \times 0.6 =$

2) $3.3 \times 0.4 =$

3) $1.28 \times 0.5 =$

4) $0.35 \times 0.6 =$

5) $1.85 \times 0.6 =$

6) $0.24 \times 0.5 =$

7) $5.25 \times 1.4 =$

8) $18.5 \times 4.6 =$

9) $15.4 \times 6.8 =$

10) $19.5 \times 2.6 =$

11) $32.2 \times 1.5 =$

12) $78.4 \times 4.5 =$

✏ **Find the quotient.**

13) $1.85 \div 10 =$

14) $74.6 \div 100 =$

15) $3.6 \div 3 =$

16) $9.6 \div 0.4 =$

17) $15.5 \div 0.5 =$

18) $32.8 \div 0.2 =$

19) $22.15 \div 1,000 =$

20) $53.55 \div 0.7 =$

21) $322.2 \div 0.2 =$

22) $50.67 \div 0.18 =$

23) $77.4 \div 0.8 =$

24) $27.93 \div 0.03 =$

Comparing Decimals

✎ **Write the correct comparison symbol (>, < or =).**

1) 0.70 ☐ 0.070

2) 0.049 ☐ 0.49

3) 5.090 ☐ 5.09

4) 2.57 ☐ 2.05

5) 9.03 ☐ 0.930

6) 6.06 ☐ 6.6

7) 7.02 ☐ 7.020

8) 3.04 ☐ 3.2

9) 3.61 ☐ 3.245

10) 0.986 ☐ 0.0986

11) 17.24 ☐ 17.240

12) 0.759 ☐ 0.81

13) 9.040 ☐ 9.40

14) 5.73 ☐ 5.213

15) 9.44 ☐ 9.404

16) 7.17 ☐ 7.170

17) 4.85 ☐ 4.085

18) 9.041 ☐ 9.40

19) 3.033 ☐ 3.030

20) 4.97 ☐ 4.970

Rounding Decimals

✎ **Round each decimal to the nearest whole number.**

1) 28.12   3) 16.22   5) 7.95

2) 6.9   4) 8.5   6) 52.7

✎ **Round each decimal to the nearest tenth.**

7) 31.761   9) 94.729   11) 13.219

8) 14.421   10) 77.89   12) 59.89

✎ **Round each decimal to the nearest hundredth.**

13) 8.428   15) 55.3786   17) 62.241

14) 23.812   16) 231.912   18) 19.447

✎ **Round each decimal to the nearest thousandth.**

19) 15.54324   21) 243.8652   23) 67.1983

20) 34.62586   22) 80.4529   24) 72.36788

Answers of Worksheets

Simplifying Fractions

1) $\frac{1}{2}$

2) $\frac{4}{5}$

3) $\frac{3}{4}$

4) $\frac{1}{2}$

5) $\frac{1}{4}$

6) $\frac{2}{3}$

7) $\frac{4}{5}$

8) $\frac{1}{4}$

9) $\frac{1}{2}$

10) $\frac{5}{7}$

11) $\frac{1}{4}$

12) $\frac{1}{2}$

13) $\frac{7}{8}$

14) $\frac{9}{10}$

15) $\frac{1}{3}$

16) $\frac{5}{14}$

17) $\frac{2}{7}$

18) $\frac{4}{27}$

19) $\frac{6}{31}$

20) $\frac{4}{9}$

21) $\frac{1}{8}$

22) C

23) A

24) B

Adding and Subtracting Fractions

1) $\frac{9}{9} = 1$

2) $\frac{9}{14}$

3) $\frac{5}{8}$

4) $1\frac{1}{10}$

5) $\frac{17}{20}$

6) $1\frac{1}{4}$

7) $1\frac{1}{5}$

8) $1\frac{1}{15}$

9) $1\frac{8}{21}$

10) $1\frac{1}{3}$

11) $1\frac{7}{30}$

12) $\frac{3}{4}$

13) $\frac{1}{6}$

14) $\frac{5}{8}$

15) $\frac{1}{6}$

16) $\frac{1}{20}$

17) $-\frac{1}{24}$

18) $\frac{3}{28}$

19) $\frac{13}{18}$

20) $\frac{7}{12}$

21) $\frac{19}{24}$

22) $-\frac{1}{15}$

23) $\frac{5}{28}$

24) $\frac{1}{2}$

25) $\frac{3}{28}$

26) $\frac{27}{40}$

27) $\frac{18}{35}$

28) $\frac{5}{16}$

29) $\frac{1}{2}$

30) $\frac{1}{18}$

Multiplying and Dividing Fractions

1) $\frac{3}{5}$

2) $\frac{1}{3}$

3) $\frac{1}{20}$

4) $\frac{1}{30}$

5) $\frac{1}{20}$

6) $\frac{1}{9}$

7) $\frac{2}{7}$

8) $\frac{1}{16}$

9) $\frac{3}{8}$

10) $\frac{1}{14}$

11) $\frac{1}{3}$

12) $\frac{1}{6}$

13) 2

14) $\frac{1}{2}$

15) $3\frac{3}{4}$

16) $\frac{2}{5}$

17) $\frac{3}{4}$ 21) $\frac{7}{10}$ 25) $\frac{5}{18}$ 29) $3\frac{3}{4}$

18) $4\frac{1}{2}$ 22) 1 26) $\frac{25}{36}$ 30) $\frac{20}{33}$

19) $\frac{26}{49}$ 23) $\frac{3}{5}$ 27) $\frac{9}{10}$

20) $\frac{2}{9}$ 24) $\frac{3}{4}$ 28) $\frac{8}{15}$

Adding and Subtracting Mixed Numbers

1) $5\frac{1}{2}$ 8) $7\frac{9}{10}$ 15) $3\frac{1}{4}$ 22) $2\frac{1}{16}$

2) 8 9) $11\frac{29}{35}$ 16) $3\frac{13}{15}$ 23) $4\frac{1}{24}$

3) $4\frac{1}{2}$ 10) $14\frac{19}{48}$ 17) $3\frac{1}{8}$ 24) $2\frac{71}{72}$

4) $4\frac{7}{12}$ 11) $1\frac{1}{2}$ 18) $2\frac{13}{15}$ 25) $5\frac{33}{35}$

5) $6\frac{5}{12}$ 12) $2\frac{1}{5}$ 19) $3\frac{7}{30}$ 26) $6\frac{32}{63}$

6) $8\frac{13}{15}$ 13) $1\frac{2}{9}$ 20) $4\frac{3}{35}$

7) $6\frac{16}{21}$ 14) $1\frac{9}{14}$ 21) $5\frac{1}{12}$

Multiplying and Dividing Mixed Numbers

1) $12\frac{3}{8}$ 10) 11 19) $4\frac{4}{27}$

2) $23\frac{1}{9}$ 11) $\frac{4}{7}$ 20) $2\frac{26}{33}$

3) $35\frac{15}{16}$ 12) $1\frac{1}{4}$ 21) $3\frac{17}{36}$

4) $8\frac{2}{3}$ 13) $4\frac{2}{9}$ 22) $\frac{19}{22}$

5) 5 14) $2\frac{23}{24}$ 23) $3\frac{27}{91}$

6) $6\frac{6}{7}$ 15) 3 24) $3\frac{13}{36}$

7) $30\frac{1}{3}$ 16) $\frac{11}{45}$ 25) $7\frac{1}{5}$

8) $7\frac{6}{7}$ 17) $\frac{19}{84}$ 26) $4\frac{16}{45}$

9) $21\frac{7}{8}$ 18) $\frac{52}{57}$

Adding and Subtracting Decimals

1) 10.91   2) 76.87   3) 94.95   4) 17.04

5) 83.93	9) 25.52	13) 3.85	17) 46.58
6) 27.89	10) 2.6	14) 6.6	18) 9.69
7) 50.95	11) 1.78	15) 26.39	19) 52.8
8) 87.17	12) 1.55	16) 6.8	

Multiplying and Dividing Decimals

1) 0.3	7) 7.35	13) 0.185	19) 0.02215
2) 1.32	8) 85.1	14) 0.746	20) 76.5
3) 0.64	9) 104.72	15) 1.2	21) 1,611
4) 0.21	10) 50.7	16) 24	22) 281.5
5) 1.11	11) 48.3	17) 31	23) 96.75
6) 0.12	12) 352.8	18) 164	24) 931

Comparing Decimals

1) >	6) <	11) =	16) =
2) <	7) =	12) <	17) >
3) =	8) <	13) <	18) <
4) >	9) >	14) >	19) >
5) >	10) >	15) >	20) =

Rounding Decimals

1) 28	9) 94.7	17) 62.24
2) 7	10) 77.9	18) 19.45
3) 16	11) 13.2	19) 15.543
4) 9	12) 59.9	20) 34.626
5) 8	13) 8.43	21) 243.865
6) 53	14) 23.81	22) 80.453
7) 31.8	15) 55.38	23) 67.198
8) 14.4	16) 231.91	24) 72.368

Chapter 3 :

Proportions, Ratios, and Percent

Topics that you'll practice in this chapter:

- ✓ Simplifying Ratios
- ✓ Proportional Ratios
- ✓ Similarity and Ratios
- ✓ Ratio and Rates Word Problems
- ✓ Percentage Calculations
- ✓ Percent Problems
- ✓ Discount, Tax and Tip
- ✓ Percent of Change
- ✓ Simple Interest

Without mathematics, there's nothing you can do. Everything around you is mathematics.
Everything around you is numbers." – Shakuntala Devi

Simplifying Ratios

✎ **Reduce each ratio.**

1) $15:20 = $ ___ : ___

2) $7:70 = $ ___ : ___

3) $16:28 = $ ___ : ___

4) $7:21 = $ ___ : ___

5) $4:40 = $ ___ : ___

6) $6:48 = $ ___ : ___

7) $16:64 = $ ___ : ___

8) $10:25 = $ ___ : ___

9) $8:48 = $ ___ : ___

10) $49:63 = $ ___ : ___

11) $18:27 = $ ___ : ___

12) $35:10 = $ ___ : ___

13) $90:9 = $ ___ : ___

14) $24:32 = $ ___ : ___

15) $7:56 = $ ___ : ___

16) $45:63 = $ ___ : ___

17) $56:72 = $ ___ : ___

18) $26:13 = $ ___ : ___

19) $15:45 = $ ___ : ___

20) $28:4 = $ ___ : ___

21) $24:48 = $ ___ : ___

22) $30:24 = $ ___ : ___

23) $70:140 = $ ___ : ___

24) $6:180 = $ ___ : ___

✎ **Write each ratio as a fraction in simplest form.**

25) $6:12 = $

26) $30:50 = $

27) $15:35 = $

28) $9:27 = $

29) $8:24 = $

30) $18:84 = $

31) $7:14 = $

32) $7:35 = $

33) $40:96 = $

34) $12:54 = $

35) $44:52 = $

36) $12:27 = $

37) $15:180 = $

38) $39:143 = $

39) $20:300 = $

40) $30:120 = $

41) $56:42 = $

42) $26:130 = $

43) $66:123 = $

44) $70:630 = $

45) $75:125 = $

Proportional Ratios

✎ **Fill in the blanks; Calculate each proportion.**

1) $3:8 = \_\_ : 48$

2) $2:5 = 20:\_\_$

3) $1:9 = \_\_ : 81$

4) $6:7 = 12:\_\_$

5) $9:2 = 63:\_\_$

6) $8:7 = \_\_ : 49$

7) $20:3 = \_\_ : 15$

8) $1:3 = \_\_ : 75$

9) $7:6 = \_\_ : 60$

10) $8:5 = \_\_ : 45$

11) $3:10 = 60:\_\_$

12) $6:11 = 42:\_\_$

✎ **State if each pair of ratios form a proportion.**

13) $\frac{3}{20}$ and $\frac{9}{60}$

14) $\frac{1}{7}$ and $\frac{6}{42}$

15) $\frac{3}{7}$ and $\frac{24}{56}$

16) $\frac{4}{9}$ and $\frac{12}{18}$

17) $\frac{1}{9}$ and $\frac{12}{81}$

18) $\frac{7}{8}$ and $\frac{21}{28}$

19) $\frac{9}{13}$ and $\frac{27}{39}$

20) $\frac{1}{8}$ and $\frac{8}{64}$

21) $\frac{6}{19}$ and $\frac{30}{85}$

22) $\frac{5}{9}$ and $\frac{40}{81}$

23) $\frac{9}{14}$ and $\frac{108}{168}$

24) $\frac{15}{23}$ and $\frac{360}{552}$

✎ **Calculate each proportion.**

25) $\frac{20}{25} = \frac{32}{x}, x = \underline{\quad}$

26) $\frac{1}{8} = \frac{32}{x}, x = \underline{\quad}$

27) $\frac{15}{5} = \frac{21}{x}, x = \underline{\quad}$

28) $\frac{1}{7} = \frac{x}{294}, x = \underline{\quad}$

29) $\frac{7}{9} = \frac{x}{81}, x = \underline{\quad}$

30) $\frac{1}{5} = \frac{13}{x}, x = \underline{\quad}$

31) $\frac{9}{5} = \frac{36}{x}, x = \underline{\quad}$

32) $\frac{6}{13} = \frac{48}{x}, x = \underline{\quad}$

33) $\frac{5}{8} = \frac{x}{88}, x = \underline{\quad}$

34) $\frac{4}{15} = \frac{x}{240}, x = \underline{\quad}$

35) $\frac{9}{19} = \frac{x}{266}, x = \underline{\quad}$

36) $\frac{7}{15} = \frac{x}{270}, x = \underline{\quad}$

Similarity and Ratios

✎ **Each pair of figures is similar. Find the missing side.**

1)

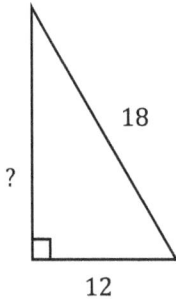

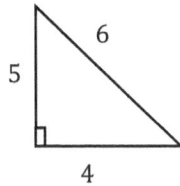

2)

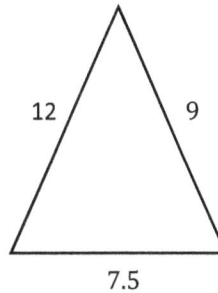

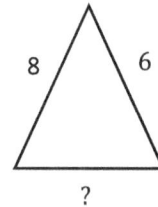

3)

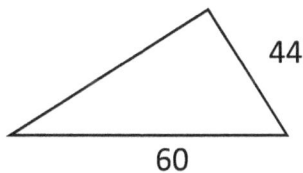

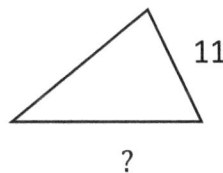

4)

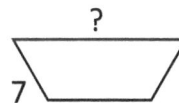

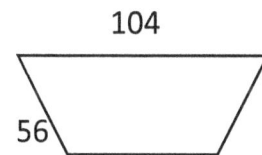

✎ **Calculate.**

5) Two rectangles are similar. The first is 24 feet wide and 120 feet long. The second is 30 feet wide. What is the length of the second rectangle? _____

6) Two rectangles are similar. One is 5 meters by 36 meters. The longer side of the second rectangle is 90 meters. What is the other side of the second rectangle? _____

7) A building casts a shadow 25 ft long. At the same time a girl 10 ft tall casts a shadow 5 ft long. How tall is the building? _____

8) The scale of a map of Texas is 4 inches: 32 miles. If you measure the distance from Dallas to Martin County as 38.4 inches, approximately how far is Martin County from Dallas? _____

Ratio and Rates Word Problems

✍ **Find the answer for each word problem.**

1) Mason has 24 red cards and 36 green cards. What is the ratio of Mason 's red cards to his green cards? _____

2) In a party, 45 soft drinks are required for every 54 guests. If there are 378 guests, how many soft drinks is required? _____

3) In Mason's class, 42 of the students are tall and 24 are short. In Michael's class 84 students are tall and 48 students are short. Which class has a higher ratio of tall to short students? _____

4) The price of 5 apples at the Quick Market is $4.6. The price of 7 of the same apples at Walmart is $5.95. Which place is the better buy? _____

5) The bakers at a Bakery can make 90 bagels in 3 hours. How many bagels can they bake in 24 hours? What is that rate per hour? _____

6) You can buy 5 cans of green beans at a supermarket for $5.75. How much does it cost to buy 45 cans of green beans? _____

7) The ratio of boys to girls in a class is 4: 7. If there are 32 boys in the class, how many girls are in that class? _____

8) The ratio of red marbles to blue marbles in a bag is 3: 7. If there are 50 marbles in the bag, how many of the marbles are red? _____

Percentage Calculations

✎ **Calculate the given percent of each value.**

1) 3% of 60 = ____

2) 20% of 32 = ____

3) 4% of 72 = ____

4) 16% of 32 = ____

5) 25% of 124 = ____

6) 35% of 56 = ____

7) 15% of 20 = ____

8) 14% of 150 = ____

9) 80% of 50 = ____

10) 12% of 115 = ____

11) 72% of 250 = ____

12) 52% of 500 = ____

13) 70% of 400 = ____

14) 27% of 145 = ____

15) 90% of 64 = ____

16) 60% of 55 = ____

17) 22% of 210 = ____

18) 8% of 235 = ____

✎ **Calculate the percent of each given value.**

19) ____% of 25 = 5

20) ____% of 40 = 20

21) ____% of 25 = 2

22) ____% of 50 = 16

23) ____% of 250 = 5

24) ____% of 40 = 32

25) ____% of 125 = 20

26) ____% of 700 = 49

27) ____% of 350 = 49

28) ____% of 500 = 210

✎ **Calculate each percent problem.**

29) A Cinema has 250 seats. 60 seats were sold for the current movie. What percent of seats are empty? _____ %

30) There are 68 boys and 92 girls in a class. 75% of the students in the class take the bus to school. How many students do not take the bus to school? ____

Percent Problems

✍ **Calculate each problem.**

1) 9 is what percent of 45? ____%

2) 60 is what percent of 120? ____%

3) 10 is what percent of 200? ____%

4) 15 is what percent of 125? ____%

5) 10 is what percent of 400? ____%

6) 66 is what percent of 55? ____%

7) 40 is what percent of 160? ____%

8) 40 is what percent of 50? ____%

9) 120 is what percent of 800? ____%

10) 78 is what percent of 120? ___%

11) 36 is what percent of 144? ___%

12) 17 is what percent of 85? ___%

13) 90 is what percent of 900? ___%

14) 36 is what percent of 16? ___%

15) 63 is what percent of 14? ___%

16) 18 is what percent of 60? ___%

17) 126 is what percent of 200? ___%

18) 232 is what percent of 40? ___%

✍ **Calculate each percent word problem.**

19) There are 40 employees in a company. On a certain day, 25 were present. What percent showed up for work? ____%

20) A metal bar weighs 60 ounces. 25% of the bar is gold. How many ounces of gold are in the bar? _____

21) A crew is made up of 12 women; the rest are men. If 15% of the crew are women, how many people are in the crew? _____

22) There are 40 students in a class and 8 of them are girls. What percent are boys? ____%

23) The Royals softball team played 400 games and won 280 of them. What percent of the games did they lose? ____%

Discount, Tax and Tip

✎ Find the selling price of each item.

1) Original price of a computer: $420
   Tax: 8% Selling price: $_____

2) Original price of a laptop: $280
   Tax: 4% Selling price: $_____

3) Original price of a sofa: $820
   Tax: 5% Selling price: $_____

4) Original price of a car: $15,800
   Tax: 3.6%   Selling price: $_____

5) Original price of a Table: $250
   Tax: 9% Selling price: $_____

6) Original price of a house: $630,000
   Tax: 1.8%   Selling price: $_____

7) Original price of a tablet: $450
   Discount: 30%   Selling price: $____

8) Original price of a chair: $390
   Discount: 8%   Selling price: $____

9) Original price of a book: $75
   Discount: 42%   Selling price: $____

10) Original price of a cellphone: $820
   Discount: 23%   Selling price: $___

11) Food bill: $45
   Tip: 15%   Price: $_____

12) Food bill: $32
   Tipp: 20%   Price: $_____

13) Food bill: $90
   Tip: 35%   Price: $_____

14) Food bill: $42
   Tipp: 12%   Price: $_____

✎ Find the answer for each word problem.

15) Nicolas hired a moving company. The company charged $500 for its services, and Nicolas gives the movers a 40% tip. How much does Nicolas tip the movers? $_____

16) Mason has lunch at a restaurant and the cost of his meal is $90. Mason wants to leave a 25% tip. What is Mason's total bill including tip? $_____

17) The sales tax in Texas is 19.80% and an item costs $350. How much is the tax? $_____

18) The price of a table at Best Buy is $680. If the sales tax is 5%, what is the final price of the table including tax? $_____

Percent of Change

✎ Find each percent of change.

1) From 150 to 450. ___ %

2) From 50 ft to 250 ft. ___ %

3) From $60 to $360. ___ %

4) From 60 cm to 180 cm. ___ %

5) From 15 to 45. ___ %

6) From 80 to 16. ___ %

7) From 120 to 360. ___ %

8) From 900 to 450. ___ %

9) From 1,000 to 200. ___ %

10) From 144 to 36. ___ %

✎ Calculate each percent of change word problem.

11) Bob got a raise, and his hourly wage increased from $42 to $63. What is the percent increase? ___ %

12) The price of a pair of shoes increases from $50 to $61. What is the percent increase? ___ %

13) At a coffee shop, the price of a cup of coffee increased from $4.80 to $5.76. What is the percent increase in the cost of the coffee? ___ %

14) 51 cm are cut from 85 cm board. What is the percent decrease in length? ___ %

15) In a class, the number of students has been increased from 54 to 81. What is the percent increase? ___ %

16) The price of gasoline rises from $24.40 to $30.50 in one month. By what percent did the gas price rise? ___ %

17) A shirt was originally priced at $38. It went on sale for $24.70. What was the percent that the shirt was discounted? ___ %

Simple Interest

✍ **Determine the simple interest for these loans.**

1) $480 at 11% for 3 years. $ _____

2) $4,200 at 7% for 4 years. $ _____

3) $2,500 at 20% for 3 years. $ _____

4) $6,800 at 3.9% for 4 months. $ ____

5) $800 at 6% for 7 months. $ _____

6) $36,000 at 4.2% for 6 years. $ _____

7) $6,500 at 7% for 4 years. $ _____

8) $850 at 9.5% for 2 years. $ _____

9) $1,200 at 5.8% for 9 months. $ ____

10) $3,000 at 4.5% for 7 years. $ _____

✍ **Calculate each simple interest word problem.**

11) A new car, valued at $22,000, depreciates at 8.5% per year. What is the value of the car one year after purchase? $_____

12) Sara puts $9,000 into an investment yielding 6% annual simple interest; she left the money in for three years. How much interest does Sara get at the end of those three years? $_____

13) A bank is offering 12% simple interest on a savings account. If you deposit $16,400, how much interest will you earn in two years? $_____

14) $720 interest is earned on a principal of $6,000 at a simple interest rate of 4% interest per year. For how many years was the principal invested? _____

15) In how many years will $2,200 yield an interest of $440 at 4% simple interest? _____

16) Jim invested $8,000 in a bond at a yearly rate of 4.5%. He earned $1,440 in interest. How long was the money invested? _____

Answers of Worksheets

Simplifying Ratios

1) $3:4$	14) $3:4$	26) $\frac{3}{5}$	36) $\frac{4}{9}$
2) $1:10$	15) $1:8$	27) $\frac{3}{7}$	37) $\frac{1}{12}$
3) $4:7$	16) $5:7$	28) $\frac{1}{3}$	38) $\frac{3}{11}$
4) $1:3$	17) $7:9$	29) $\frac{1}{3}$	39) $\frac{1}{15}$
5) $1:10$	18) $2:1$	30) $\frac{3}{14}$	40) $\frac{1}{4}$
6) $1:8$	19) $1:3$	31) $\frac{1}{2}$	41) $\frac{4}{3}$
7) $2:8$	20) $7:1$	32) $\frac{1}{5}$	42) $\frac{1}{5}$
8) $2:5$	21) $1:2$	33) $\frac{5}{12}$	43) $\frac{22}{41}$
9) $1:6$	22) $5:4$	34) $\frac{2}{9}$	44) $\frac{1}{9}$
10) $7:9$	23) $1:2$	35) $\frac{11}{13}$	45) $\frac{3}{5}$
11) $2:3$	24) $1:30$		
12) $7:2$	25) $\frac{1}{2}$		
13) $10:1$			

Proportional Ratios

1) 18	10) 72	19) Yes	28) 42
2) 50	11) 200	20) Yes	29) 63
3) 9	12) 77	21) No	30) 65
4) 14	13) Yes	22) No	31) 20
5) 14	14) Yes	23) Yes	32) 104
6) 56	15) Yes	24) Yes	33) 55
7) 100	16) No	25) 40	34) 64
8) 25	17) No	26) 256	35) 126
9) 70	18) No	27) 7	36) 126

Similarity and ratios

1) 15	4) 13	7) 50 feet
2) 5	5) 150 feet	8) 307.2 miles
3) 15	6) 12.5 meters	

Ratio and Rates Word Problems

1) $2:3$	2) 315

3) The ratio for both classes is 7 to 4.   6) $51.75

4) Walmart is a better buy.   7) 56

5) 720, the rate is 30 per hour.   8) 15

Percentage Calculations

1) 1.8	11) 180	21) 8%
2) 6.4	12) 260	22) 32%
3) 2.88	13) 280	23) 2%
4) 5.12	14) 39.15	24) 80%
5) 31	15) 57.6	25) 16%
6) 19.6	16) 33	26) 7%
7) 3	17) 46.2	27) 14%
8) 21	18) 18.8	28) 42%
9) 40	19) 20%	29) 76%
10) 13.8	20) 50%	30) 40

Percent Problems

1) 20%	9) 15%	17) 63%
2) 50%	10) 65%	18) 580%
3) 5%	11) 25%	19) 62.5%
4) 12%	12) 20%	20) 15 ounces
5) 2.5%	13) 10%	21) 80
6) 120%	14) 225%	22) 80%
7) 25%	15) 450%	23) 30%
8) 80%	16) 30%	

Discount, Tax and Tip

1) $453.60	7) $315.00	13) $121.50
2) $291.20	8) $358.80	14) $47.04
3) $861.00	9) $43.50	15) $200.00
4) $16,368.80	10) $631.40	16) $112.50
5) $272.50	11) $51.75	17) $69.30
6) $641,340	12) $38.40	18) $714.00

Percent of Change

1) 200%	7) 200%	13) 20%
2) 400%	8) 50%	14) 60%
3) 500%	9) 80%	15) 50%
4) 200%	10) 75%	16) 25%
5) 200%	11) 50%	17) 35%
6) 80%	12) 22%	

Simple Interest

1) $158.40	7) $1,820.00	13) $3,936.00
2) $1,176.00	8) $161.50	14) 3 years
3) $1,500.00	9) $52.20	15) 5 years
4) $88.40	10) $945.00	16) 4 years
5) $28.00	11) $20,130.00	
6) $9,072.00	12) $1,620.00	

Chapter 4 :

Exponents and Radicals Expressions

Topics that you'll practice in this chapter:

✓ Multiplication Property of Exponents

✓ Zero and Negative Exponents

✓ Division Property of Exponents

✓ Powers of Products and Quotients

✓ Negative Exponents and Negative Bases

✓ Scientific Notation

✓ Square Roots

✓ Simplifying Radical Expressions

✓ Simplifying Radical Expressions Involving Fractions

✓ Multiplying Radical Expressions

✓ Adding and Subtracting Radical Expressions

Love is anterior to life, posterior to death, initial of creation, and the exponent of breath.

Emily Dickinson

Multiplication Property of Exponents

✏ **Simplify and write the answer in exponential form.**

1) $4 \times 4^5 =$

2) $8^4 \times 8 =$

3) $7^3 \times 7^3 =$

4) $9^2 \times 9^2 =$

5) $2^2 \times 2^4 \times 2 =$

6) $5 \times 5^3 \times 5^3 =$

7) $4^3 \times 4^2 \times 4 \times 4 =$

8) $5x \times x =$

9) $x^3 \times x^3 =$

10) $x^7 \times x^2 =$

11) $x^4 \times x^3 \times x^2 =$

12) $10x \times 3x =$

13) $4x^3 \times 4x^3 =$

14) $7x^3 \times x =$

15) $3x^2 \times 4x^2 \times x^2 =$

16) $5x^4 \times x^4 =$

17) $2x^8 \times 2x =$

18) $6x \times x^5 =$

19) $4x^2 \times 6x^6 =$

20) $5yx^3 \times 4x =$

21) $7x^3 \times y^5 x^7 =$

22) $y^2 x^3 \times y^5 x^4 =$

23) $3x^5 \times 4x^3 y^4 =$

24) $4x^4 \times 9x^2 y^5 =$

25) $5x^3 y^4 \times 6x^8 y^2 =$

26) $8x^3 y^6 \times 4xy^3 =$

27) $2xy^5 \times 6x^3 y^3 =$

28) $4x^5 y^2 \times 4x^2 y^8 =$

29) $7x \times 3y^8 x^2 \times y^5 =$

30) $x^3 \times 2y^3 x^4 \times 2y =$

31) $3yx^4 \times 3y^4 x \times 3xy^3 =$

32) $6y^3 \times 2y^2 x^4 \times 10yx^5 =$

Zero and Negative Exponents

✎ **Evaluate the following expressions.**

1) $1^{-5} =$

2) $4^{-1} =$

3) $0^{10} =$

4) $1^{15} =$

5) $5^{-2} =$

6) $3^{-3} =$

7) $9^{-1} =$

8) $10^{-2} =$

9) $12^{-2} =$

10) $2^{-5} =$

11) $3^{-4} =$

12) $2^{-4} =$

13) $6^{-3} =$

14) $10^{-3} =$

15) $30^{-1} =$

16) $15^{-2} =$

17) $4^{-3} =$

18) $2^{-7} =$

19) $5^{-3} =$

20) $4^{-4} =$

21) $3^{-5} =$

22) $10^{-4} =$

23) $2^{-10} =$

24) $8^{-3} =$

25) $20^{-2} =$

26) $14^{-2} =$

27) $9^{-3} =$

28) $100^{-2} =$

29) $5^{-4} =$

30) $4^{-6} =$

31) $\left(\frac{1}{4}\right)^{-3}$

32) $\left(\frac{1}{6}\right)^{-2} =$

33) $\left(\frac{1}{7}\right)^{-2} =$

34) $\left(\frac{2}{3}\right)^{-3} =$

35) $\left(\frac{1}{13}\right)^{-2} =$

36) $\left(\frac{7}{12}\right)^{-2} =$

37) $\left(\frac{1}{6}\right)^{-3} =$

38) $\left(\frac{1}{300}\right)^{-2} =$

39) $\left(\frac{2}{9}\right)^{-2} =$

40) $\left(\frac{7}{5}\right)^{-1} =$

41) $\left(\frac{13}{23}\right)^{0} =$

42) $\left(\frac{1}{4}\right)^{-5} =$

Division Property of Exponents

✏ **Simplify.**

1) $\dfrac{5^6}{5^7} =$

2) $\dfrac{8^8}{8^6} =$

3) $\dfrac{4^5}{4} =$

4) $\dfrac{3}{3^5} =$

5) $\dfrac{x}{x^6} =$

6) $\dfrac{3 \times 3^2}{3^2 \times 3^5} =$

7) $\dfrac{9^4}{9^2} =$

8) $\dfrac{10 \times 10^9}{10^2 \times 10^7} =$

9) $\dfrac{7^5 \times 7^7}{7^4 \times 7^8} =$

10) $\dfrac{15x}{30x^6} =$

11) $\dfrac{3x^9}{4x^4} =$

12) $\dfrac{15x^8}{10x^9} =$

13) $\dfrac{42x^5}{6y^9} =$

14) $\dfrac{36y^8}{4x^4y^5} =$

15) $\dfrac{2x^7}{9x} =$

16) $\dfrac{49x^8y^6}{7x^9} =$

17) $\dfrac{48x^2}{24x^6y^{12}} =$

18) $\dfrac{30yx^5}{6yx^7} =$

19) $\dfrac{19x^7y}{38x^{12}y^4} =$

20) $\dfrac{9x^8}{63x^8} =$

21) $\dfrac{9x^{-9}}{4x^{-3}} =$

Powers of Products and Quotients

🖎 **Simplify.**

1) $(4^3)^2 =$

2) $(2^3)^4 =$

3) $(2 \times 2^3)^2 =$

4) $(5 \times 5^5)^6 =$

5) $(19^4 \times 19^2)^3 =$

6) $(2^3 \times 2^4)^4 =$

7) $(5 \times 5^2)^2 =$

8) $(4^4)^4 =$

9) $(8x^5)^2 =$

10) $(3x^2y^4)^4 =$

11) $(7x^5y^2)^2 =$

12) $(5x^4y^4)^3 =$

13) $(2x^3y^3)^5 =$

14) $(10x^3y^4)^3 =$

15) $(13y^3y)^2 =$

16) $(5x^6x^4)^2 =$

17) $(6x^7y^6)^3 =$

18) $(12x^5x^7)^2 =$

19) $(2x^4 \times 2x)^4 =$

20) $(2x^4y^3)^5 =$

21) $(15x^7y^2)^2 =$

22) $(8x^3y^5)^3 =$

23) $(3x \times 2y^2)^4 =$

24) $\left(\frac{4x}{x^5}\right)^2 =$

25) $\left(\frac{x^4y^5}{x^3y^5}\right)^9 =$

26) $\left(\frac{36xy}{6x^5}\right)^3 =$

27) $\left(\frac{x^7}{x^8y^2}\right)^6 =$

28) $\left(\frac{xy^4}{x^3y^6}\right)^{-3} =$

29) $\left(\frac{5xy^8}{x^3}\right)^2 =$

30) $\left(\frac{xy^6}{2xy^3}\right)^{-4} =$

Negative Exponents and Negative Bases

✍ **Simplify.**

1) $-9^{-1} =$

2) $-9^{-2} =$

3) $-2^{-5} =$

4) $-x^{-7} =$

5) $11x^{-1} =$

6) $-8x^{-3} =$

7) $-12x^{-5} =$

8) $-9x^{-8}y^{-6} =$

9) $32x^{-5}y^{-1} =$

10) $10a^{-9}b^{-3} =$

11) $-17x^4y^{-6} =$

12) $-\dfrac{25}{x^{-5}} =$

13) $-\dfrac{13x}{a^{-7}} =$

14) $\left(-\dfrac{1}{3}\right)^{-4} =$

15) $\left(-\dfrac{3}{4}\right)^{-2} =$

16) $-\dfrac{14}{a^{-6}b^{-3}} =$

17) $-\dfrac{7x}{x^{-8}} =$

18) $-\dfrac{a^{-9}}{b^{-5}} =$

19) $-\dfrac{11}{x^{-5}} =$

20) $\dfrac{8b}{-16c^{-6}} =$

21) $\dfrac{12ab}{a^{-4}b^{-3}} =$

22) $-\dfrac{8n^{-4}}{32p^{-7}} =$

23) $\dfrac{16ab^{-6}}{-6c^{-5}} =$

24) $\left(\dfrac{10a}{5c}\right)^{-4} =$

25) $\left(-\dfrac{12x}{4yz}\right)^{-3} =$

26) $\dfrac{8ab^{-7}}{-5c^{-3}} =$

27) $\left(-\dfrac{x^4}{x^5}\right)^{-5} =$

28) $\left(-\dfrac{x^{-2}}{7x^3}\right)^{-2} =$

29) $\left(-\dfrac{x^{-4}}{x^2}\right)^{-6} =$

Scientific Notation

✎ **Write each number in scientific notation.**

1) 0.223 =

2) 0.09 =

3) 4.5 =

4) 900 =

5) 2,000 =

6) 0.006 =

7) 33 =

8) 9,400 =

9) 1,470 =

10) 52,000 =

11) 8,000,000 =

12) 0.00009 =

13) 2,158,000 =

14) 0.0039 =

15) 0.000075 =

16) 4,300,000 =

17) 130,000 =

18) 4,000,000,000 =

19) 0.00009 =

20) 0.0039 =

✎ **Write each number in standard notation.**

21) 4×10^{-1} =

22) 1.2×10^{-3} =

23) 2.7×10^5 =

24) 6×10^{-4} =

25) 3.6×10^{-3} =

26) 5.5×10^5 =

27) 3.2×10^4 =

28) 3.88×10^6 =

29) 7×10^{-6} =

30) 4.2×10^{-7} =

Square Roots

✎ **Find the value each square root.**

1) $\sqrt{16} = $ ___

2) $\sqrt{25} = $ ___

3) $\sqrt{1} = $ ___

4) $\sqrt{64} = $ ___

5) $\sqrt{0} = $ ___

6) $\sqrt{196} = $ ___

7) $\sqrt{4} = $ ___

8) $\sqrt{256} = $ ___

9) $\sqrt{36} = $ ___

10) $\sqrt{289} = $ ___

11) $\sqrt{169} = $ ___

12) $\sqrt{144} = $ ___

13) $\sqrt{100} = $ ___

14) $\sqrt{1,600} = $ ___

15) $\sqrt{2,500} = $ ___

16) $\sqrt{324} = $ ___

17) $\sqrt{529} = $ ___

18) $\sqrt{20} = $ ___

19) $\sqrt{625} = $ ___

20) $\sqrt{18} = $ ___

21) $\sqrt{50} = $ ___

22) $\sqrt{1,024} = $ ___

23) $\sqrt{160} = $ ___

24) $\sqrt{32} = $ ___

✎ **Evaluate.**

25) $\sqrt{4} \times \sqrt{25} = $ _____

26) $\sqrt{36} \times \sqrt{49} = $ _____

27) $\sqrt{6} \times \sqrt{6} = $ _____

28) $\sqrt{13} \times \sqrt{13} = $ _____

29) $2\sqrt{5} \times 3\sqrt{5} = $ _____

30) $\sqrt{12} \times \sqrt{3} = $ _____

31) $\sqrt{13} + \sqrt{13} = $ _____

32) $\sqrt{10} + 2\sqrt{10} = $ _____

33) $12\sqrt{7} - 10\sqrt{7} = $ _____

34) $4\sqrt{10} \times 2\sqrt{10} = $ _____

35) $5\sqrt{3} \times 8\sqrt{3} = $ _____

36) $6\sqrt{3} - \sqrt{12} = $ _____

Simplifying Radical Expressions

✍ **Simplify.**

1) $\sqrt{13x^2} =$

2) $\sqrt{75x^2} =$

3) $\sqrt[3]{27a} =$

4) $\sqrt{64x^5} =$

5) $\sqrt{216a} =$

6) $\sqrt[3]{63w^3} =$

7) $\sqrt{192x} =$

8) $\sqrt{125v} =$

9) $\sqrt[3]{128x^2} =$

10) $\sqrt{100x^9} =$

11) $\sqrt{16x^4} =$

12) $\sqrt[3]{500a^5} =$

13) $\sqrt{242} =$

14) $\sqrt{392p^3} =$

15) $\sqrt{8m^6} =$

16) $\sqrt{198x^3y^3} =$

17) $\sqrt{121x^5y^5} =$

18) $\sqrt{16a^6b^3} =$

19) $\sqrt{90x^5y^7} =$

20) $\sqrt[3]{64y^2x^6} =$

21) $10\sqrt{16x^4} =$

22) $6\sqrt{81x^2} =$

23) $\sqrt[3]{56x^2y^6} =$

24) $\sqrt[3]{1,000x^5y^7} =$

25) $8\sqrt{50a} =$

26) $\sqrt[4]{625x^8y} =$

27) $\sqrt{24x^4y^5r^3} =$

28) $5\sqrt{36x^4y^5z^8} =$

29) $3\sqrt[3]{343x^9y^7} =$

30) $5\sqrt{81a^5b^2c^9} =$

31) $\sqrt[4]{625x^8y^{16}} =$

Multiplying Radical Expressions

✎ **Simplify.**

1) $\sqrt{5} \times \sqrt{5} =$

2) $\sqrt{5} \times \sqrt{10} =$

3) $\sqrt{3} \times \sqrt{12} =$

4) $\sqrt{49} \times \sqrt{47} =$

5) $\sqrt{7} \times -2\sqrt{28} =$

6) $3\sqrt{15} \times \sqrt{5} =$

7) $4\sqrt{72} \times \sqrt{2} =$

8) $\sqrt{5} \times -\sqrt{49} =$

9) $\sqrt{55} \times \sqrt{11} =$

10) $7\sqrt{42} \times 2\sqrt{216} =$

11) $\sqrt{45}(5 + \sqrt{5}) =$

12) $\sqrt{13x^2} \times \sqrt{13x^3} =$

13) $-2\sqrt{27} \times \sqrt{3} =$

14) $2\sqrt{13x^4} \times \sqrt{13x^4} =$

15) $\sqrt{14x^3} \times \sqrt{7x^2} =$

16) $-8\sqrt{5x} \times \sqrt{7x^5} =$

17) $-2\sqrt{16x^5} \times 4\sqrt{8x^3} =$

18) $-4\sqrt{32}(8 + \sqrt{32}) =$

19) $\sqrt{32x}(10 - \sqrt{2x}) =$

20) $\sqrt{2x}(8\sqrt{x^5} + \sqrt{8}) =$

21) $\sqrt{20r}(5 + \sqrt{5}) =$

22) $-4\sqrt{7x} \times 3\sqrt{14x^5} =$

23) $-2\sqrt{12x} \times 3\sqrt{2x}$

24) $-\sqrt{7v^3}(-3\sqrt{42v}) =$

25) $(\sqrt{11} - 5)(\sqrt{11} + 5) =$

26) $(-3\sqrt{5} + 3)(\sqrt{5} - 4) =$

27) $(4 - 6\sqrt{3})(-6 + \sqrt{3}) =$

28) $(8 - 3\sqrt{5})(7 - \sqrt{5}) =$

29) $(-1 - \sqrt{3x})(4 + \sqrt{3x}) =$

30) $(-5 + 2\sqrt{7r})(-5 + \sqrt{7r}) =$

31) $(-\sqrt{7n} + 1)(-\sqrt{7} - 5) =$

32) $(-3 + \sqrt{3})(5 - 2\sqrt{3x}) =$

Simplifying Radical Expressions Involving Fractions

✎ **Simplify.**

1) $\dfrac{\sqrt{5}}{\sqrt{3}} =$

2) $\dfrac{\sqrt{18}}{\sqrt{45}} =$

3) $\dfrac{\sqrt{10}}{5\sqrt{2}} =$

4) $\dfrac{13}{\sqrt{3}} =$

5) $\dfrac{12\sqrt{5r}}{\sqrt{m^5}} =$

6) $\dfrac{11\sqrt{2}}{\sqrt{k}} =$

7) $\dfrac{6\sqrt{20x^3}}{\sqrt{16x}} =$

8) $\dfrac{\sqrt{14x^3y^4}}{\sqrt{7x^4y^3}} =$

9) $\dfrac{1}{1-\sqrt{5}} =$

10) $\dfrac{1-8\sqrt{a}}{\sqrt{11a}} =$

11) $\dfrac{\sqrt{a}}{\sqrt{a}+\sqrt{b}} =$

12) $\dfrac{1-\sqrt{5}}{2-\sqrt{6}} =$

13) $\dfrac{4+\sqrt{7}}{3-\sqrt{8}} =$

14) $\dfrac{5}{-3-3\sqrt{3}} =$

15) $\dfrac{7}{2-\sqrt{5}} =$

16) $\dfrac{\sqrt{7}-\sqrt{3}}{\sqrt{3}-\sqrt{7}} =$

17) $\dfrac{\sqrt{5}+\sqrt{7}}{\sqrt{7}-\sqrt{5}} =$

18) $\dfrac{2\sqrt{2}-\sqrt{3}}{3\sqrt{2}+\sqrt{5}} =$

19) $\dfrac{\sqrt{11}+5\sqrt{3}}{4-\sqrt{11}} =$

20) $\dfrac{\sqrt{5}+\sqrt{3}}{2-\sqrt{3}} =$

21) $\dfrac{\sqrt{32a^7b^4}}{\sqrt{2ab^3}} =$

22) $\dfrac{10\sqrt{21x^5}}{5\sqrt{x^3}} =$

Adding and Subtracting Radical Expressions

✍ **Simplify.**

1) $\sqrt{2} + \sqrt{8} =$

2) $3\sqrt{50} + 4\sqrt{2} =$

3) $2\sqrt{12} - 4\sqrt{3} =$

4) $5\sqrt{32} - 5\sqrt{2} =$

5) $3\sqrt{75} - 5\sqrt{3} =$

6) $-\sqrt{72} - 4\sqrt{2} =$

7) $-7\sqrt{16} - 4\sqrt{25} =$

8) $8\sqrt{24} + 2\sqrt{6} =$

9) $10\sqrt{49} - 7\sqrt{100} =$

10) $-7\sqrt{5} + 9\sqrt{45} =$

11) $-15\sqrt{12} + 14\sqrt{48} =$

12) $20\sqrt{4} - 2\sqrt{25} =$

13) $-2\sqrt{20} + 7\sqrt{5} =$

14) $8\sqrt{7} - 2\sqrt{63} =$

15) $5\sqrt{44} + 3\sqrt{11} =$

16) $3\sqrt{27} - 5\sqrt{48} =$

17) $\sqrt{144} - \sqrt{81} =$

18) $3\sqrt{20} - 6\sqrt{5} =$

19) $-2\sqrt{7} + 8\sqrt{28} =$

20) $3\sqrt{75} - 2\sqrt{3} =$

21) $5\sqrt{27} - 3\sqrt{3} =$

22) $-7\sqrt{30} + 6\sqrt{120} =$

23) $-7\sqrt{24} - 2\sqrt{6} =$

24) $-\sqrt{32x} + 4\sqrt{2x} =$

25) $\sqrt{7y^2} + y\sqrt{112} =$

26) $\sqrt{45mn^2} + 2n\sqrt{5m} =$

27) $-4\sqrt{12a} - 4\sqrt{3a} =$

28) $-5\sqrt{15ab} - 2\sqrt{60ab} =$

29) $\sqrt{45x^2y} + x\sqrt{20y} =$

30) $2\sqrt{7a} + 4\sqrt{63a} =$

Answers of Worksheets

Multiplication Property of Exponents

1) 4^6
2) 8^5
3) 7^6
4) 9^4
5) 2^7
6) 5^7
7) 4^7
8) $5x^2$

9) x^6
10) x^9
11) x^9
12) $30x^2$
13) $16x^6$
14) $7x^4$
15) $12x^6$
16) $5x^8$

17) $4x^9$
18) $6x^6$
19) $24x^8$
20) $20x^4y$
21) $7x^{10}y^5$
22) x^7y^7
23) $12x^8y^4$
24) $36x^6y^5$

25) $30x^{11}y^6$
26) $32x^4y^9$
27) $12x^4y^8$
28) $16x^7y^{10}$
29) $21x^3y^{13}$
30) $4x^7y^4$
31) $27x^6y^8$
32) $120x^9y^6$

Zero and Negative Exponents

1) 1
2) $\frac{1}{4}$
3) 0
4) 1
5) $\frac{1}{25}$
6) $\frac{1}{27}$
7) $\frac{1}{9}$
8) $\frac{1}{100}$
9) $\frac{1}{144}$
10) $\frac{1}{32}$
11) $\frac{1}{81}$

12) $\frac{1}{16}$
13) $\frac{1}{216}$
14) $\frac{1}{1,000}$
15) $\frac{1}{30}$
16) $\frac{1}{225}$
17) $\frac{1}{64}$
18) $\frac{1}{128}$
19) $\frac{1}{125}$
20) $\frac{1}{256}$
21) $\frac{1}{243}$

22) $\frac{1}{10,000}$
23) $\frac{1}{1,024}$
24) $\frac{1}{512}$
25) $\frac{1}{400}$
26) $\frac{1}{196}$
27) $\frac{1}{729}$
28) $\frac{1}{10,000}$
29) $\frac{1}{625}$
30) $\frac{1}{4,096}$
31) 64
32) 36

33) 49
34) $\frac{27}{8}$
35) 169
36) $\frac{144}{49}$
37) 216
38) $90,000$
39) $\frac{81}{4}$
40) $\frac{5}{7}$
41) 1
42) $1,024$

Division Property of Exponents

1) $\frac{1}{5}$
2) 8^2
3) 4^4
4) $\frac{1}{3^4}$

5) $\frac{1}{x^5}$
6) $\frac{1}{3^4}$
7) 9^2
8) 10

9) 1
10) $\frac{1}{2x^5}$
11) $\frac{3x^5}{4}$
12) $\frac{3}{2x}$

13) $\frac{7x^5}{y^9}$
14) $\frac{9y^3}{x^4}$
15) $\frac{2x^6}{9}$

16) $\frac{7y^6}{x}$ 18) $\frac{5}{x^2}$ 19) $\frac{1}{2x^5y^3}$ 21) $\frac{9}{4x^6}$

17) $\frac{2}{x^4y^{12}}$ 20) $\frac{1}{7}$

Powers of Products and Quotients

1) 4^6

2) 2^{12}

3) 2^8

4) 5^{36}

5) 19^{18}

6) 2^{28}

7) 5^6

8) 4^{16}

9) $64x^{10}$

10) $81x^8y^{16}$

11) $49x^{10}y^4$

12) $125x^{12}y^{12}$

13) $32x^{15}y^{15}$

14) $1,000x^9y^{12}$

15) $169y^8$

16) $25x^{20}$

17) $216x^{21}y^{18}$

18) $144x^{24}$

19) $256x^{20}$

20) $32x^{20}y^{15}$

21) $225x^{14}y^4$

22) $512x^9y^{15}$

23) $1,296x^4y^8$

24) $\frac{16}{x^8}$

25) x^9

26) $\frac{216y^3}{x^{12}}$

27) $\frac{1}{x^6y^{12}}$

28) x^6y^6

29) $\frac{25y^{16}}{x^4}$

30) $\frac{16}{y^{12}}$

Negative Exponents and Negative Bases

1) $-\frac{1}{9}$

2) $-\frac{1}{81}$

3) $-\frac{1}{32}$

4) $-\frac{1}{x^7}$

5) $\frac{11}{x}$

6) $-\frac{8}{x^3}$

7) $-\frac{12}{x^5}$

8) $-\frac{9}{x^8y^6}$

9) $\frac{32}{x^5y}$

10) $\frac{10}{a^9b^3}$

11) $-\frac{17x^4}{y^6}$

12) $-25x^5$

13) $-13xa^7$

14) 81

15) $\frac{16}{9}$

16) $-14a^6b^3$

17) $-7x^9$

18) $-\frac{b^5}{a^9}$

19) $-11x^5$

20) $-\frac{bc^6}{2}$

21) $12a^5b^4$

22) $-\frac{p^7}{4n^4}$

23) $-\frac{8ac^5}{3b^6}$

24) $\frac{c^4}{16a^4}$

25) $\frac{y^3z^3}{27x^3}$

26) $-\frac{8ac^3}{5b^7}$

27) $-x^5$

28) $49x^{10}$

29) x^{36}

Scientific Notation

1) 2.23×10^{-1}
2) 9×10^{-2}
3) 4.5×10^{0}
4) 9×10^{2}
5) 2×10^{3}
6) 6×10^{-3}
7) 3.3×10^{1}
8) 9.4×10^{3}
9) 1.47×10^{3}
10) 5.2×10^{4}

11) 8×10^{6}
12) 9×10^{-5}
13) 2.158×10^{6}
14) 3.9×10^{-3}
15) 7.5×10^{-5}
16) 4.3×10^{6}
17) 1.3×10^{5}
18) 4×10^{9}
19) 9×10^{-5}
20) 3.9×10^{-3}

21) 0.4
22) 0.0012
23) 270,000
24) 0.0006
25) 0.0036
26) 550,000
27) 32,000
28) 3,880,000
29) 0.000007
30) 0.00000042

Square Roots

1) 4
2) 5
3) 1
4) 8
5) 0
6) 14
7) 2
8) 16
9) 6

10) 17
11) 13
12) 12
13) 10
14) 40
15) 50
16) 18
17) 23
18) $2\sqrt{5}$

19) 25
20) $3\sqrt{2}$
21) $5\sqrt{2}$
22) 32
23) $4\sqrt{10}$
24) $4\sqrt{2}$
25) 10
26) 42
27) 6

28) 13
29) 30
30) 6
31) $2\sqrt{13}$
32) $3\sqrt{10}$
33) $2\sqrt{7}$
34) 80
35) 120
36) $4\sqrt{3}$

Simplifying radical expressions

1) $x\sqrt{13}$
2) $5x\sqrt{3}$
3) $3\sqrt[3]{a}$
4) $8x^2\sqrt{x}$
5) $6\sqrt{6a}$
6) $w\sqrt[3]{63}$
7) $8\sqrt{3x}$
8) $5\sqrt{5v}$

9) $4\sqrt[3]{2x^2}$
10) $10x^4\sqrt{x}$
11) $4x^2$
12) $5a\sqrt[3]{4a^2}$
13) $11\sqrt{2}$
14) $14p\sqrt{2p}$
15) $2m^3\sqrt{2}$
16) $3x.y\sqrt{22xy}$

17) $11x^2y^2\sqrt{xy}$
18) $4a^3b\sqrt{b}$
19) $3x^2y^3\sqrt{10xy}$
20) $4x^2\sqrt[3]{y^2}$
21) $40x^2$
22) $54x$
23) $2y^2\sqrt[3]{7x^2}$
24) $10xy^2\sqrt[3]{x^2y}$

25) $40\sqrt{2a}$ 28) $30x^2y^2z^4\sqrt{y}$ 31) $5x^2y^4$

26) $5x^2\sqrt[4]{y}$ 29) $21x^3y^2\sqrt[3]{y}$

27) $2x^2y^2r\sqrt{6yr}$ 30) $45a^2bc^4\sqrt{ac}$

Multiplying radical expressions

1) 5

2) $5\sqrt{2}$

3) 6

4) $7\sqrt{47}$

5) -28

6) $15\sqrt{3}$

7) 48

8) $-5\sqrt{7}$

9) $11\sqrt{5}$

10) $504\sqrt{7}$

11) $15\sqrt{5} + 15$

12) $13x^2\sqrt{x}$

13) -18

14) $26x^4$

15) $7x^2\sqrt{2x}$

16) $-8x^3\sqrt{35}$

17) $-64x^4\sqrt{2}$

18) $-128\sqrt{2} - 128$

19) $40\sqrt{2x} - 8x$

20) $8x^3\sqrt{2} + 4\sqrt{x}$

21) $10\sqrt{5r} + 10\sqrt{r}$

22) $-84x^3\sqrt{2}$

23) $-12\sqrt{6}x$

24) $21v^2\sqrt{6}$

25) -14

26) $15\sqrt{5} - 27$

27) $40\sqrt{3} - 42$

28) $71 - 29\sqrt{5}$

29) $-3x - 5\sqrt{3x} - 4$

30) $14r - 15\sqrt{7r} + 25$

31) $7\sqrt{n} + 5\sqrt{7n} - \sqrt{7} - 5$

32) $-15 + 6\sqrt{3x} + 5\sqrt{3} - 6\sqrt{x}$

Simplifying radical expressions involving fractions

1) $\frac{\sqrt{15}}{3}$

2) $\frac{9\sqrt{10}}{45} = \frac{\sqrt{10}}{5}$

3) $\frac{\sqrt{20}}{10} = \frac{\sqrt{5}}{5}$

4) $\frac{13\sqrt{3}}{3}$

5) $\frac{12\sqrt{5mr}}{m^3}$

6) $\frac{11\sqrt{2k}}{k}$

7) $3x\sqrt{5}$

8) $\frac{\sqrt{2x}}{xy}$

9) $\frac{-1-\sqrt{5}}{4}$

10) $\frac{\sqrt{11a}-8a\sqrt{11}}{11a}$

11) $\frac{a-\sqrt{ab}}{a-b}$

12) $\frac{\sqrt{30}+2\sqrt{5}-\sqrt{6}-2}{2}$

13) $12 + 8\sqrt{2} + 3\sqrt{7} + 2\sqrt{14}$

14) $-\frac{5(\sqrt{3}-1)}{6}$

15) $-14 - 7\sqrt{5}$

16) -1

17) $6 + \sqrt{35}$

18) $\frac{12 - 2\sqrt{10} - 3\sqrt{6} + \sqrt{15}}{13}$

19) $\frac{4\sqrt{11}+11+20\sqrt{3}+5\sqrt{33}}{5}$

20) $2\sqrt{5} + 3 + \sqrt{15} + 2\sqrt{3}$

21) $4a^3\sqrt{b}$

22) $2x\sqrt{21}$

Adding and subtracting radical expressions

1) $3\sqrt{2}$

2) $19\sqrt{2}$

3) 0

4) $15\sqrt{2}$

5) $10\sqrt{3}$

6) $-10\sqrt{2}$

7) -48

8) $18\sqrt{6}$

9) 0

10) $20\sqrt{5}$

11) $26\sqrt{3}$

12) 30

13) $3\sqrt{5}$

14) $2\sqrt{7}$

15) $13\sqrt{11}$

16) $-11\sqrt{3}$

17) 3

18) 0

19) $14\sqrt{7}$

20) $13\sqrt{3}$

21) $12\sqrt{3}$

22) $5\sqrt{30}$

23) $-16\sqrt{6}$

24) 0

25) $5y\sqrt{7}$

26) $5n\sqrt{5m}$

27) $-12\sqrt{3a}$

28) $-9\sqrt{15ab}$

29) $5x\sqrt{5y}$

30) $14\sqrt{7a}$

Chapter 5 :

Algebraic Expressions

Topics that you'll practice in this chapter:

- ✓ Simplifying Variable Expressions
- ✓ Simplifying Polynomial Expressions
- ✓ Translate Phrases into an Algebraic Statement
- ✓ The Distributive Property
- ✓ Evaluating One Variable Expressions
- ✓ Evaluating Two Variables Expressions
- ✓ Combining like Terms

I want freedom for the full expression of my personality.
Mahatma Gandhi

Simplifying Variable Expressions

✎ **Simplify each expression.**

1) $3(x + 5) =$

2) $(-4)(7x - 5) =$

3) $11x + 5 - 6x =$

4) $-4 - 2x^2 - 6x^2 =$

5) $7 + 13x^2 + 3 =$

6) $3x^2 + 7x + 15x^2 =$

7) $3x^2 - 12x^2 + 4x =$

8) $4x^2 - 8x - 2x =$

9) $6x + 7(3 - 4x) =$

10) $8x + 4(15x - 3) =$

11) $6(-3x - 9) - 17 =$

12) $-11x^2 - (-5x) =$

13) $2x + 7 + 5 - 8x =$

14) $7 + 6x - 11 - 5x =$

15) $27x + 8 - 13 - 5x =$

16) $(-11)(-5x + 2) - 41x =$

17) $19x - 4(4 - 2x) =$

18) $16x + 3(3x + 6) + 10 =$

19) $5(-2x - 4) - 13x =$

20) $16x - 3x(x + 10) =$

21) $17x + 5x(2 - 4x) =$

22) $5x(-4x - 7) + 20x =$

23) $25x - 19 + 4x^2 =$

24) $6x(x - 11) + 25 =$

25) $4x - 5 + 15x + 3x^2 =$

26) $-7x^2 - 11x - 9x =$

27) $10x - 9x^2 - 3x^2 - 7 =$

28) $13 + 3x^2 - 9x^2 - 21x =$

29) $22x + 10x^2 - 15x + 17 =$

30) $4x^2 + 25x + 21x^2 =$

31) $29 - 12x^2 - 23x - 4x^2 =$

32) $22x - 19x - 9x^2 + 30 =$

Simplifying Polynomial Expressions

✎ **Simplify each polynomial.**

1) $(2x^3 + 8x^2) - (11x + 3x^2) =$ _____

2) $(2x^5 + 7x^3) - (5x^3 + 11x^2) =$ _____

3) $(41x^4 + 5x^2) - (4x^2 + 20x^4) =$ _____

4) $13x - 8x^2 + 4(4x^2 + 3x^3) =$ _____

5) $(4x^3 - 22) + 5(3x^2 - 6x^3) =$ _____

6) $(4x^3 - 3x) - 5(2x^3 + x^4) =$ _____

7) $5(5x - 2x^3) - 2(8x^3 + 5x^2) =$ _____

8) $(3x^2 - 10x) - (5x^3 + 14x^2) =$ _____

9) $5x^3 - (3x^4 + 5x) + 2x^2 =$ _____

10) $11x^4 - (3x^2 + 5x) + 7x =$ _____

11) $(6x^2 - 3x^4) - (10x^4 + 3x^2) =$ _____

12) $2x^2 - 7x^3 + 19x^4 - 22x^3 =$ _____

13) $10x^2 - x^4 + 4x^4 - 32x^3 =$ _____

14) $-5x^2 + 17x^3 - 8x^2 - 6x =$ _____

15) $x^4 - 11x^5 - 30x^4 + 5x^2 =$ _____

16) $21x^3 + 13x - 5x^2 - 11x^3 =$ _____

Translate Phrases into an Algebraic Statement

✎ Write an algebraic expression for each phrase.

1) 9 multiplied by x. _____

2) Subtract 11 from y. _____

3) 19 divided by x. _____

4) 38 decreased by y. _____

5) Add y to 40. _____

6) The square of 6. _____

7) x raised to the fifth power. _____

8) The sum of six and a number. _____

9) The difference between fifty–seven and y. _____

10) The quotient of nine and a number. _____

11) The quotient of the square of x and 25. _____

12) The difference between x and 6 is 19. _____

13) 10 times a reduced by the square of b. _____

14) Subtract the product of a and b from 41. _____

The Distributive Property

✍ **Use the distributive property to simply each expression.**

1) $4(1 + 2x) =$

2) $2(4 + 7x) =$

3) $3(4x - 4) =$

4) $(2x - 5)(-6) =$

5) $(-3)(x + 6) =$

6) $(4 + 3x)2 =$

7) $(-5)(8 - 3x) =$

8) $-(-5 - 7x) =$

9) $(-6x + 3)(-3) =$

10) $(-4)(x - 7) =$

11) $-(5 - 3x) =$

12) $3(9 + 4x) =$

13) $6(4 + 3x) =$

14) $(-5x + 3)2 =$

15) $(5 - 8x)(-3) =$

16) $(-12)(3x + 3) =$

17) $(5 - 3x)6 =$

18) $4(2 + 6x) =$

19) $8(7x - 3) =$

20) $(-2x + 3)4 =$

21) $(7 - 5x)(-9) =$

22) $(-10)(x - 8) =$

23) $(11 - 4x)3 =$

24) $(-6)(10x - 4) =$

25) $(3 - 9x)(-7) =$

26) $(-9)(x + 9) =$

27) $(-3 + 5x)(-7) =$

28) $(-5)(8 - 10x) =$

29) $12(4x - 8) =$

30) $(-10x + 13)(-3) =$

31) $(-8)(3x - 2) + 4(x + 5) =$

32) $(-8)(x + 4) - (6 + 5x) =$

Evaluating One Variable Expressions

✎ **Evaluate each expression using the value given.**

1) $8 - x$, $x = 5$

2) $x - 9$, $x = 5$

3) $5x + 4$, $x = 3$

4) $x - 13$, $x = -4$

5) $12 - x$, $x = 4$

6) $x + 2$, $x = 6$

7) $4x + 8$, $x = 3$

8) $x + (-7)$, $x = -8$

9) $4x + 5$, $x = 2$

10) $3x + 9$, $x = -2$

11) $15 + 3x - 7$, $x = 2$

12) $17 - 3x$, $x = 3$

13) $8x - 9$, $x = 4$

14) $5x + 4$, $x = -3$

15) $10x + 5$, $x = 3$

16) $14 - 4x$, $x = -6$

17) $3(5x + 3)$, $x = 9$

18) $4(-3x - 6)$, $x = 3$

19) $7x - 2x + 12$, $x = 4$

20) $(5x + 6) \div 2$, $x = 8$

21) $(x + 18) \div 10$, $x = 12$

22) $5x - 12 + 3x$, $x = -3$

23) $(6 - 4x)(-3)$, $x = -4$

24) $9x^2 + 3x - 6$, $x = 2$

25) $x^2 - 10x$, $x = -5$

26) $3x(7 - 2x)$, $x = 2$

27) $12x + 6 - 2x^2$, $x = -4$

28) $(-3)(4x - 8 + 3x)$, $x = 3$

29) $(-6) + \frac{x}{4} + 3x$, $x = 16$

30) $(-6) + \frac{x}{5}$, $x = 35$

31) $\left(-\frac{45}{x}\right) - 7 + 2x$, $x = 9$

32) $\left(-\frac{21}{x}\right) - 12 + 4x$, $x = 7$

Evaluating Two Variables Expressions

✎ Evaluate each expression using the values given.

1) $2x - 4y$,

   $x = 4, y = 1$

2) $3x + 5y$,

   $x = -2, y = 2$

3) $-7a + 4b$,

   $a = 2, b = 4$

4) $3x + 5 - y$,

   $x = 5, y = 6$

5) $3z + 12 - 2k$,

   $z = 5, k = 6$

6) $6(-x - 3y)$,

   $x = 5, y = -2$

7) $5a + 3b$,

   $a = 3, b = 4$

8) $7x \div 3y$,

   $x = 3, y = 7$

9) $2x + 15 + 5y$,

   $x = -3, y = 1$

10) $5a - (18 - b)$,

   $a = 2, b = 8$

11) $2z + 20 + 5k$,

   $z = -6, k = 5$

12) $xy + 10 + 4x$,

   $x = 3, y = 5$

13) $2x + 4y - 8 + 5$,

   $x = 5, y = 2$

14) $\left(-\frac{24}{x}\right) + 3 + 2y$,

   $x = 4, y = 6$

15) $(-3)(-3a - 3b)$,

   $a = 4, b = 5$

16) $12 + 4x - 7 - y$,

   $x = 3, y = 5$

17) $11x + 5 - 8y + 6$,

   $x = 5, y = 2$

18) $10 + 2(-4x - 5y)$,

   $x = 5, y = 4$

19) $5x + 13 + 6y$,

   $x = 5, y = 6$

20) $10a - (7a + 3b) - 11$,

   $a = 3, b = 8$

Combining like Terms

✎ **Simplify each expression.**

1) $11x + 3x + 6 =$

2) $8(2x - 6) =$

3) $18x - 7x + 11 =$

4) $(-4)(6x - 7) =$

5) $22x - 10x - 5 =$

6) $32x - 13 + 8x =$

7) $15 - (8x - 11) =$

8) $-24x + 17 - 11x =$

9) $12x - 8 - 6x + 9 =$

10) $21x + 5 - 36 + 12x =$

11) $28x + 3x - 11 =$

12) $(-3x + 4)5 =$

13) $2 + 4x + 9x - 8 =$

14) $6(2x - 5x) - 4 =$

15) $4(5x + 11) + 3x =$

16) $x - 14 - 11x =$

17) $5(10 + 9x) - 8x =$

18) $42x + 17 - 23x =$

19) $(-7x) + 19 + 20x =$

20) $(-7x) - 33 + 29x =$

21) $4(5x + 3) - 19x =$

22) $5(6 - 2x) - 15x =$

23) $-24x + (11 - 18x) =$

24) $(-9) - (6)(7x + 3) =$

25) $(-1)(8x - 10) - 21x =$

26) $-36x + 14 + 27x - 5x =$

27) $3(-13x + 6) - 17x =$

28) $-5x - 42 + 32x =$

29) $37x - 19x + 15 - 9x =$

30) $3(5x + 7x) - 31 =$

31) $14 - 6x - 15 - 9x =$

32) $-2(-5x - 7x) + 27x =$

Answers of Worksheets

Simplifying Variable Expressions

1) $3x + 15$

2) $-28x + 20$

3) $5x + 5$

4) $-8x^2 - 4$

5) $13x^2 + 10$

6) $18x^2 + 7x$

7) $-9x^2 + 4x$

8) $4x^2 - 10x$

9) $-22x + 21$

10) $68x - 12$

11) $-18x - 71$

12) $-11x^2 + 5x$

13) $-6x + 12$

14) $x - 4$

15) $22x - 5$

16) $14x - 22$

17) $27x - 16$

18) $25x + 28$

19) $-23x - 20$

20) $-3x^2 - 14x$

21) $-20x^2 + 27x$

22) $-20x^2 - 15x$

23) $4x^2 + 25x - 19$

24) $6x^2 - 66x + 25$

25) $3x^2 + 19x - 5$

26) $-7x^2 - 20x$

27) $-12x^2 + 10x - 7$

28) $-6x^2 - 21x + 13$

29) $10x^2 + 7x + 17$

30) $25x^2 + 25x$

31) $-16x^2 - 23x + 29$

32) $-9x^2 + 3x + 30$

Simplifying Polynomial Expressions

1) $2x^3 + 5x^2 - 11x$

2) $2x^5 + 2x^3 - 11x^2$

3) $21x^4 + x^2$

4) $12x^3 + 8x^2 + 13x$

5) $-26x^3 + 15x^2 - 22$

6) $-5x^4 - 6x^3 - 3x$

7) $-26x^3 - 10x^2 + 25x$

8) $-5x^3 - 11x^2 - 10x$

9) $-3x^4 + 5x^3 + 2x^2 - 5x$

10) $11x^4 - 3x^2 + 2x$

11) $-13x^4 + 3x^2$

12) $19x^4 - 29x^3 + 2x^2$

13) $3x^4 - 32x^3 + 10x^2$

14) $17x^3 - 13x^2 - 6x$

15) $-11x^5 - 29x^4 + 5x^2$

16) $10x^3 - 5x^2 + 13x$

Translate Phrases into an Algebraic Statement

1) $9x$

2) $y - 11$

3) $\frac{19}{x}$

4) $38 - y$

5) $y + 40$

6) 6^2

7) x^5

8) $6 + x$

9) $57 - y$

10) $\frac{9}{x}$

11) $\frac{x^2}{25}$

12) $x - 6 = 19$

13) $10a - b^2$

14) $41 - ab$

The Distributive Property

1) $8x + 4$

2) $14x + 8$

3) $12x - 12$

4) $-12x + 30$

5) $-3x - 18$

6) $6x + 8$

7) $15x - 40$

8) $7x + 5$

9) $18x - 9$

10) $-4x + 28$

11) $3x - 5$

12) $12x + 27$

13) $18x + 24$

14) $-10x + 6$

15) $24x - 15$

16) $-36x - 36$

17) $-18x + 30$

18) $24x + 8$

19) $56x - 24$

20) $-8x + 12$

21) $45x - 63$

22) $-10x + 80$

23) $-12x + 33$

24) $-60x + 24$

25) $63x - 21$

26) $-9x - 81$

27) $-35x + 21$

28) $50x - 40$

29) $48x - 96$

30) $30x - 39$

31) $-20x + 36$

32) $-13x - 38$

Evaluating One Variables

1) 3

2) −4

3) 19

4) −17

5) 8

6) 8

7) 20

8) − 15

9) 13

10) 3

11) 14

12) 8

13) 23

14) −11

15) 35

16) 38

17) 144

18) −60

19) 32

20) 23

21) 3

22) −36

23) −66

24) 36

25) 75

26) 18

27) −74

28) −39

29) 46

30) 1

31) 6

32) 13

Evaluating Two Variables

1) 4

2) 4

3) 2

4) 14

5) 15

6) 6

7) 27

8) 1

9) 14

10) 0

11) 33

12) 37

13) 15

14) 9

15) 81

16) 12

17) 50

18) −70

19) 74

20) −26

Combining like Terms

1) $14x + 6$

2) $16x - 48$

3) $11x + 11$

4) $-24x + 28$

5) $12x - 5$

6) $40x - 13$

7) $-8x + 26$

8) $-35x + 17$

9) $6x + 1$

10) $33x - 31$

11) $31x - 11$

12) $-15x + 20$

13) $13x - 6$

14) $-18x - 4$

15) $23x + 44$

16) $-10x - 14$

17) $37x + 50$

18) $19x + 17$

19) $13x + 19$

20) $22x - 33$

21) $x + 12$

22) $-25x + 30$

23) $-42x + 11$

24) $-42x - 27$

25) $-29x + 10$

26) $-14x + 14$

27) $-56x + 18$

28) $27x - 42$

29) $9x + 15$

30) $36x - 31$

31) $-15x - 1$

32) $51x$

Chapter 6 :

Equations and Inequalities

Topics that you'll practice in this chapter:

- ✓ One–Step Equations
- ✓ Multi–Step Equations
- ✓ Graphing Single–Variable Inequalities
- ✓ One–Step Inequalities
- ✓ Multi-Step Inequalities
- ✓ Systems of Equations
- ✓ Systems of Equations Word Problems

"Life is a math equation. In order to gain the most, you have to know how to convert negatives into positives." – Anonymous

One–Step Equations

✍ **Find the answer for each equation.**

1) $3x = 90, x =$ ____

2) $5x = 35, x =$ ____

3) $6x = 24, x =$ ____

4) $24x = 144, x =$ ____

5) $x + 15 = 20, x =$ ____

6) $x - 7 = 4, x =$ ____

7) $x - 9 = 2, x =$ ____

8) $x + 15 = 23, x =$ ____

9) $x - 4 = 13, x =$ ____

10) $12 = 16 + x, x =$ ____

11) $x - 10 = 2, x =$ ____

12) $5 - x = -11, x =$ ____

13) $28 = -6 + x, x =$ ____

14) $x - 20 = -35, x =$ ____

15) $x + 14 = -4, x =$ ____

16) $14 = 28 - x, x =$ ____

17) $7 + x = -7, x =$ ____

18) $x - 16 = 4, x =$ ____

19) $30 = x - 15, x =$ ____

20) $x - 5 = -18, x =$ ____

21) $x - 10 = 24, x =$ ____

22) $x - 20 = -25, x =$ ____

23) $x - 17 = 30, x =$ ____

24) $-70 = x - 28, x =$ ____

25) $x - 9 = 13, x =$ ____

26) $36 = 4x, x =$ ____

27) $x - 35 = 25, x =$ ____

28) $x - 25 = 10, x =$ ____

29) $70 - x = 16, x =$ ____

30) $x - 10 = 14, x =$ ____

31) $17 - x = -13, x =$ __

32) $x - 9 = -30, x =$ ____

Multi–Step Equations

✎ **Find the answer for each equation.**

1) $3x + 3 = 9$

2) $-x + 5 = 12$

3) $4x - 8 = 8$

4) $-(3 - x) = 5$

5) $4x - 8 = 16$

6) $12x - 15 = 9$

7) $2x - 18 = 2$

8) $4x + 8 = 16$

9) $24x + 27 = 75$

10) $-14(3 + x) = 14$

11) $-3(2 + x) = 6$

12) $12 = -(x - 7)$

13) $3(3 - x) = 30$

14) $-15 = -(3x + 6)$

15) $40(3 + x) = 40$

16) $5(x - 10) = 25$

17) $-18 = x + 8x$

18) $3x + 25 = -2x - 10$

19) $7(6 + 3x) = -63$

20) $18 - 3x = -4 - 5x$

21) $4 - 6x = 36 + 2x$

22) $15 + 15x = -5 + 5x$

23) $42 = (-6x) - 7 + 7$

24) $21 = 3x - 21 + 4x$

25) $-18 = -6x - 9 + 3x$

26) $5x - 15 = -29 + 6x$

27) $7x - 18 = 4x + 3$

28) $-7 - 4x = 5(4 - x)$

29) $x - 5 = -5(-3 - x)$

30) $13x - 68 = 15x - 102$

31) $-5x - 3 = -3(9 + 3x)$

32) $-2x - 15 = 6x + 17$

Graphing Single–Variable Inequalities

✎ **Draw a graph for each inequality.**

1) $x > -1$

2) $x \leq 2$

3) $x \geq 0$

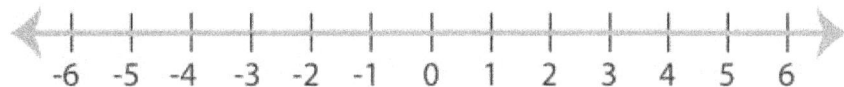

4) $x < -3$

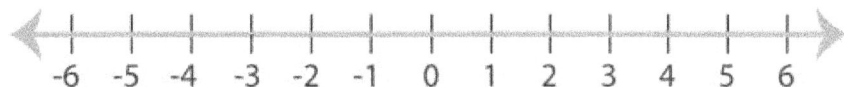

5) $x < \frac{1}{2}$

6) $x \leq -2$

7) $x \leq 3$

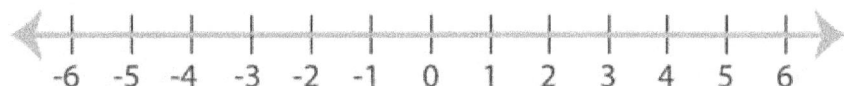

8) $x \geq -\frac{7}{2}$

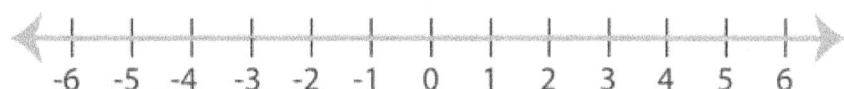

One–Step Inequalities

✍ **Find the answer for each inequality and graph it.**

1) $x + 4 \geq 4$

2) $x - 5 \leq 2$

3) $5x > 35$

4) $9 + x \leq 11$

5) $x - 5 < -9$

6) $9x \geq 72$

7) $9x \leq 27$

8) $x + 19 > 16$

Multi-Step Inequalities

✎ **Calculate each inequality.**

1) $x - 3 \leq 7$

2) $8 - x \leq 8$

3) $3x - 9 \leq 9$

4) $4x - 4 \geq 8$

5) $x - 7 \geq 1$

6) $5x - 15 \leq 5$

7) $6x - 8 \leq 4$

8) $-11 + 6x \leq 12$

9) $4(x - 4) \leq 16$

10) $3x - 10 \leq 11$

11) $5x - 25 < 25$

12) $9x - 5 < 22$

13) $20 - 7x \geq -15$

14) $33 + 6x < 45$

15) $8 + 8x \geq 96$

16) $7 + 3x < 13$

17) $4x - 3 < 9$

18) $5(2 - 2x) \geq -30$

19) $-(7 + 6x) < 29$

20) $12 - 8x \geq -20$

21) $-4(x - 6) > 24$

22) $\dfrac{3x + 9}{6} \leq 10$

23) $\dfrac{4x - 10}{3} \leq 2$

24) $\dfrac{2x - 8}{3} > 2$

25) $8 + \dfrac{x}{6} < 9$

26) $\dfrac{9x}{7} - 4 < 5$

27) $\dfrac{15x + 45}{15} > 1$

28) $16 + \dfrac{x}{4} < 6$

Systems of Equations

✎ **Calculate each system of equations.**

1) $-x + y = 2$ $x =$ ___

    $-4x + 2y = 6$ $y =$ ___

2) $-15x + 3y = -9$ $x =$ ___

    $9x - 16y = 48$ $y =$ ___

3) $y = -7$ $x =$ ___

    $6x + 5y = 7$ $y =$ ___

4) $3y = -9x + 15$ $x =$ ___

    $5x - 4y = -3$ $y =$ ___

5) $10x - 9y = -13$ $x =$ ___

    $-5x + 3y = 11$ $y =$ ___

6) $-12x - 16y = 20$ $x =$ ___

    $6x - 12y = 30$ $y =$ ___

7) $5x - 14y = -23$ $x =$ ___

    $-18x + 21y = 24$ $y =$ ___

8) $15x - 21y = -6$ $x =$ ___

    $2x - 3y = -2$ $y =$ ___

9) $-x + 3y = 3$ $x =$ ___

    $-14x + 16y = -10$ $y =$ ___

10) $x + 5y = 50$ $x =$ ___

    $3x + 10y = 80$ $y =$ ___

11) $6x - 7y = -8$ $x =$ ___

    $-x - 4y = -9$ $y =$ ___

12) $2x + 4y = -10$ $x =$ ___

    $2x - 8y = 14$ $y =$ ___

13) $4x + 3y = 12$ $x =$ ___

    $5x - 3y = 15$ $y =$ ___

14) $3x - 2y = 3$ $x =$ ___

    $7x - 8y = 22$ $y =$ ___

15) $3x + 2y = 5$ $x =$ ___

    $-10x - 4y = -14$ $y =$ ___

16) $10x + 7y = 1$ $x =$ ___

    $-5x - 7y = 24$ $y =$ ___

Systems of Equations Word Problems

✎ **Find the answer for each word problem.**

1) Tickets to a movie cost $4 for adults and $3 for students. A group of friends purchased 8 tickets for $31.00. How many adults ticket did they buy? ____

2) At a store, Eva bought two shirts and five hats for $77.00. Nicole bought three same shirts and four same hats for $84.00. What is the price of each shirt? _____

3) A farmhouse shelters 18 animals, some are pigs, and some are ducks. Altogether there are 66 legs. How many pigs are there? _____

4) A class of 214 students went on a field trip. They took 36 vehicles, some cars and some buses. If each car holds 5 students and each bus hold 22 students, how many buses did they take? _____

5) A theater is selling tickets for a performance. Mr. Smith purchased 5 senior tickets and 3 child tickets for $105 for his friends and family. Mr. Jackson purchased 3 senior tickets and 5 child tickets for $79. What is the price of a senior ticket? $_____

6) The difference of two numbers is 10. Their sum is 20. What is the bigger number? $_____

7) The sum of the digits of a certain two–digit number is 7. Reversing its digits increase the number by 9. What is the number? _____

8) The difference of two numbers is 11. Their sum is 25. What are the numbers? _____

9) The length of a rectangle is 5 meters greater than 2 times the width. The perimeter of rectangle is 28 meters. What is the length of the rectangle? _____

10) Jim has 25 nickels and dimes totaling $1.80. How many nickels does he have? _____

Answers of Worksheets

One–Step Equations

1) 30	9) 17	17) −14	25) 22
2) 7	10) −4	18) 20	26) 9
3) 4	11) 12	19) 45	27) 60
4) 6	12) 16	20) −13	28) 35
5) 5	13) 34	21) 34	29) 54
6) 11	14) −15	22) −5	30) 24
7) 11	15) −18	23) 47	31) 30
8) 8	16) 14	24) −42	32) −21

Multi–Step Equations

1) 2	9) 2	17) −2	25) 3
2) −7	10) −4	18) −7	26) 14
3) 4	11) −4	19) −5	27) 7
4) 8	12) −5	20) −11	28) 27
5) 6	13) −7	21) −4	29) −5
6) 2	14) 3	22) −2	30) 17
7) 10	15) −2	23) −7	31) −6
8) 2	16) 15	24) 6	32) −4

Graphing Single–Variable Inequalities

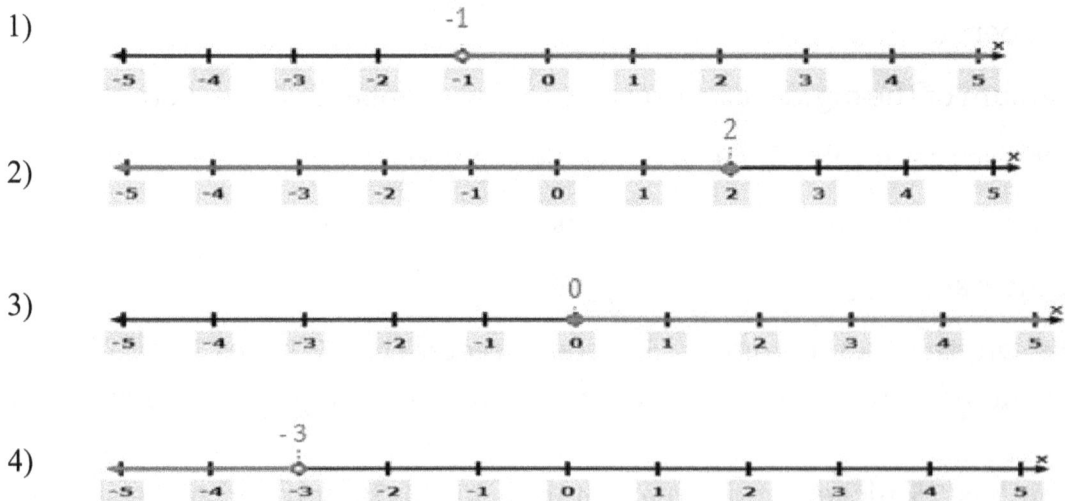

1)

2)

3)

4)

5)

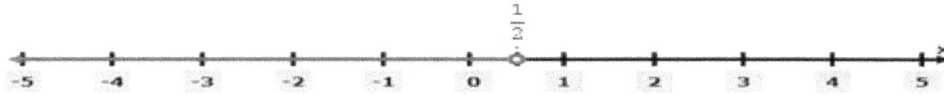

6)

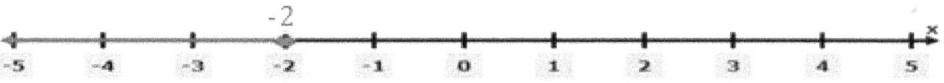

7)

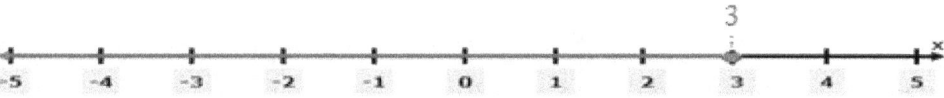

8)

One–Step Inequalities

1)

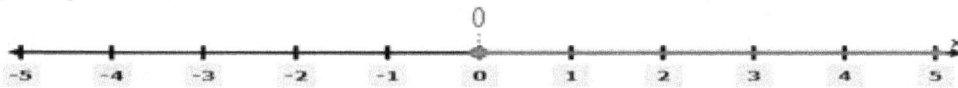

2)

3)

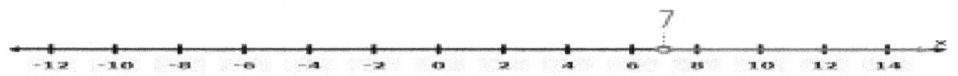

4)

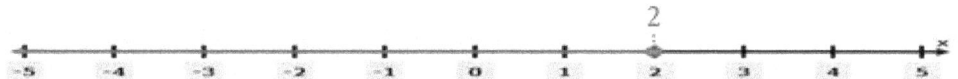

5)

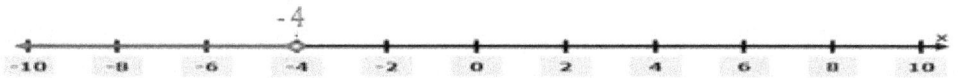

6)

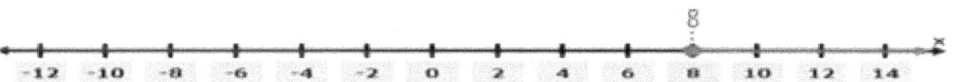

7)

8)

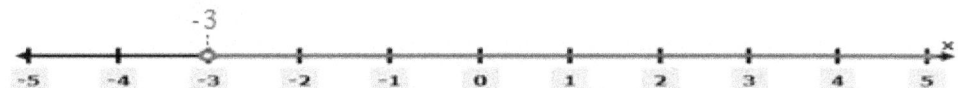

Multi-Step Inequalities

1) $x \le 10$	7) $x \le 2$	13) $x \le 5$	19) $x > -6$
2) $x \ge 0$	8) $x \le \frac{23}{6}$	14) $x < 2$	20) $x \le 4$
3) $x \le 6$	9) $x \le 8$	15) $x \ge 11$	21) $x < 0$
4) $x \ge 3$	10) $x \le 7$	16) $x < 2$	22) $x \le 17$
5) $x \ge 8$	11) $x < 10$	17) $x < 3$	23) $x \le 4$
6) $x \le 4$	12) $x < 3$	18) $x \le 4$	24) $x > 7$

25) $x < 6$ 26) $x < 7$ 27) $x > -2$ 28) $x < -40$

Systems of Equations

1) $x = -1, y = 1$ 7) $x = 1, y = 2$ 13) $x = 3, y = 0$

2) $x = 0, y = -3$ 8) $x = 8, y = 6$ 14) $x = -2, y = -\frac{9}{2}$

3) $x = 7$ 9) $x = 3, y = 2$ 15) $x = 1, y = 1$

4) $x = 1, y = 2$ 10) $x = -20, y = 14$ 16) $x = 5, y = -7$

5) $x = -4, y = -3$ 11) $x = 1, y = 2$

6) $x = 1, y = -2$ 12) $x = -1, y = -2$

Systems of Equations Word Problems

1) 7   5) $18   9) 11 meters

2) $16   6) 15   10) 14

3) 15   7) 34

4) 2   8) 18, 7

Chapter 7 :

Linear Functions

Topics that you'll practice in this chapter:

- ✓ Finding Slope
- ✓ Graphing Lines Using Line Equation
- ✓ Writing Linear Equations
- ✓ Graphing Linear Inequalities
- ✓ Finding Midpoint
- ✓ Finding Distance of Two Points

Life is not linear; you have ups and downs. It's how you deal with the troughs that defines you.

Michael Lee-Chin

Finding Slope

✒ **Find the slope of each line.**

1) $y = x + 8$

2) $y = -3x + 5$

3) $y = 2x + 12$

4) $y = -4x + 19$

5) $y = 11 + 6x$

6) $y = 7 - 5x$

7) $y = 8x + 19$

8) $y = -9x + 20$

9) $y = -7x + 4$

10) $y = 3x - 8$

11) $y = \frac{1}{3}x + 8$

12) $y = -\frac{4}{5}x + 9$

13) $-3x + 6y = 30$

14) $4x + 4y = 16$

15) $3y - x = 10$

16) $8y - x = 5$

✒ **Find the slope of the line through each pair of points.**

17) $(2, 3), (7, 10)$

18) $(-3, 5), (2, 15)$

19) $(5, -3), (1, 9)$

20) $(-5, -5), (10, 25)$

21) $(22, 3), (7, 18)$

22) $(-16, 8), (-7, 26)$

23) $(25, 11), (29, 19)$

24) $(26, -19), (14, 17)$

25) $(22, -13), (20, -11)$

26) $(19, 7), (15, -3)$

27) $(5, 7), (11, 19)$

28) $(52, -62), (40, 70)$

Graphing Lines Using Line Equation

✎ **Sketch the graph of each line.**

1) $y = x - 2$

2) $y = -3x + 2$

3) $x + y = 0$

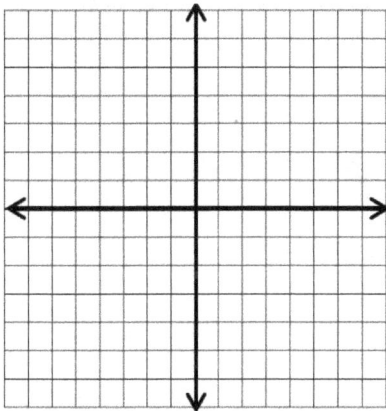 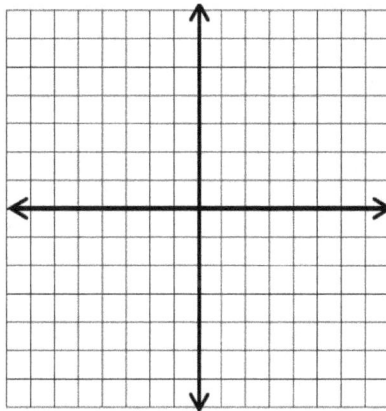 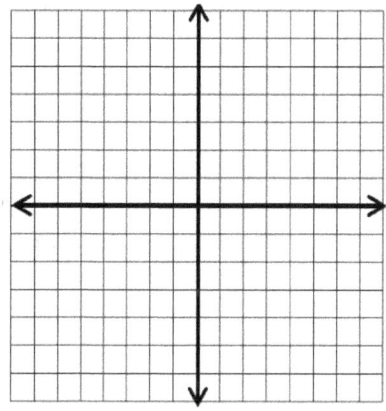

4) $x + y = -3$

5) $2x + 3y = -4$

6) $y - 3x + 6 = 0$

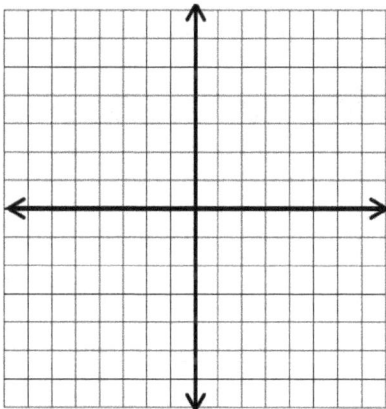 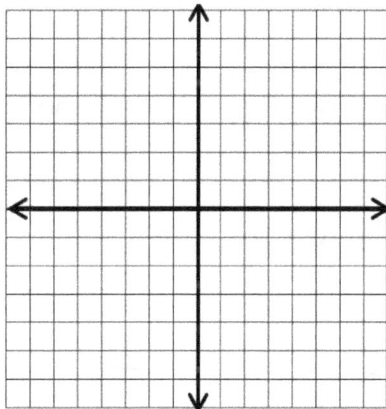 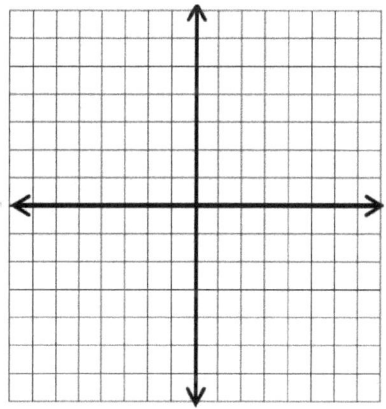

Writing Linear Equations

✍ **Write the equation of the line through the given points.**

1) Through: $(2, -5), (3, 9)$

2) Through: $(-6, 3), (3, 12)$

3) Through: $(10, 7), (5, 27)$

4) Through: $(15, 11), (3, -1)$

5) Through: $(24, 17), (12, -7)$

6) Through: $(8, 29), (4, -7)$

7) Through: $(20, -16), (12, 0)$

8) Through: $(-3, 10), (2, -5)$

9) Through: $(-6, 17), (4, -3)$

10) Through: $(-8, 22), (5, -4)$

11) Through: $(9, 27), (3, -3)$

12) Through: $(11, 32), (9, 4)$

13) Through: $(-3, 13), (-4, 0)$

14) Through: $(-5, 5), (5, 15)$

15) Through: $(18, -32), (11, 3)$

16) Through: $(-4, 25), (4, -15)$

✍ **Find the answer for each problem.**

17) What is the equation of a line with slope 6 and intercept 12?

_____

18) What is the equation of a line with slope -11 and intercept-4?

_____

19) What is the equation of a line with slope -3 and passes through point $(5, 2)$? _____

20) What is the equation of a line with slope -5 and passes through point $(-2, -1)$? _____

21) The slope of a line is -10 and it passes through point $(-3, 0)$. What is the equation of the line? _____

22) The slope of a line is 8 and it passes through point $(0, 7)$. What is the equation of the line? _____

Graphing Linear Inequalities

✏ **Sketch the graph of each linear inequality.**

1) $y > 4x - 5$ 2) $y < 2x + 4$ 3) $y \leq -5x - 2$

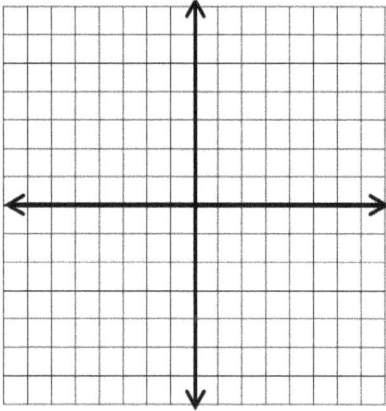 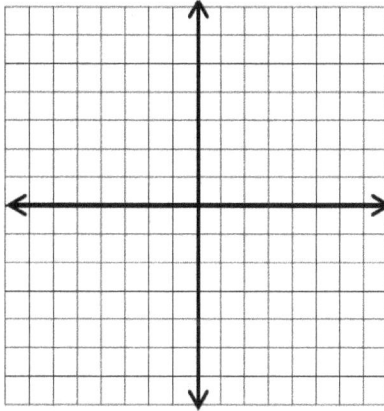 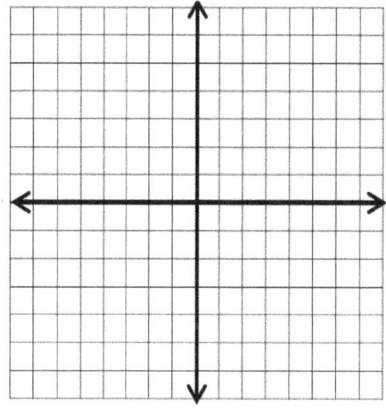

4) $4y \geq 12 + 4x$ 5) $-12y < 3x - 24$ 6) $5y \geq -15x + 10$

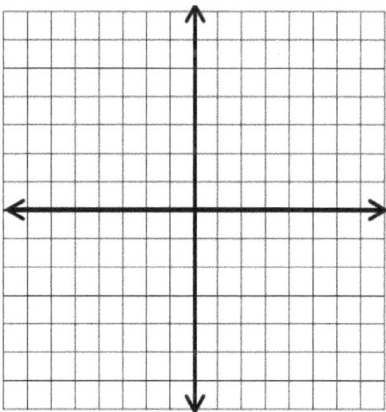 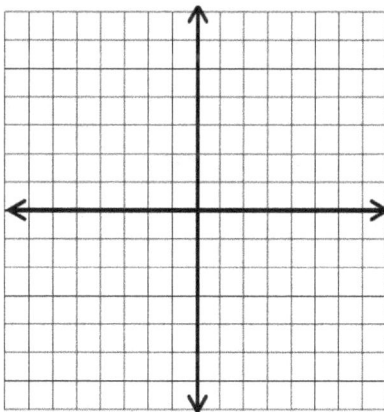 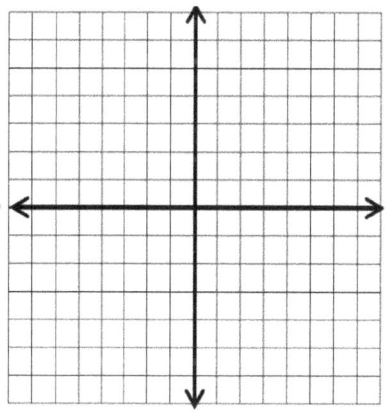

Finding Midpoint

✎ **Find the midpoint of the line segment with the given endpoints.**

1) $(-4, -3), (2, 3)$

2) $(9, 0), (-1, 8)$

3) $(9, -6), (3, 14)$

4) $(-10, -6), (0, 8)$

5) $(2, -5), (14, -15)$

6) $(-10, -3), (4, -13)$

7) $(8, 7), (-8, 13)$

8) $(-3, 6), (-9, 2)$

9) $(-4, 5), (16, -9)$

10) $(7, 14), (9, -2)$

11) $(-8, 6), (6, 6)$

12) $(10, 5), (-2, -3)$

13) $(-5, 12), (-3, 3)$

14) $(12, 7), (8, -2)$

15) $(10, 2), (-6, 14)$

16) $(-1, -2), (-7, 10)$

17) $(7, -7), (13, -13)$

18) $(-3, -8), (11, -4)$

19) $(5, -11), (-8, 9)$

20) $(14, -4), (16, 14)$

21) $(0, -5), (8, -1)$

22) $(3, 0), (-21, 18)$

23) $(17, -3), (-7, -5)$

24) $(26, -12), (6, 24)$

✎ **Find the answer for each problem.**

25) One endpoint of a line segment is $(-3, 7)$ and the midpoint of the line segment is $(-6, 9)$. What is the other endpoint? _____

26) One endpoint of a line segment is $(-3, 7)$ and the midpoint of the line segment is $(1, 5)$. What is the other endpoint? _____

27) One endpoint of a line segment is $(-10, -16)$ and the midpoint of the line segment is $(2, 9)$. What is the other endpoint? _____

Finding Distance of Two Points

✍ **Find the distance between each pair of points.**

1) $(6, 3), (-3, -9)$

2) $(5, 2), (-10, -6)$

3) $(8, 5), (8, 3)$

4) $(-8, -2), (2, 22)$

5) $(6, -7), (-3, -7)$

6) $(12, 0), (-9, -20)$

7) $(3, 20), (3, -5)$

8) $(10, 17), (5, 5)$

9) $(7, -2), (-4, -2)$

10) $(13, 4), (5, -2)$

11) $(11, 13), (5, 5)$

12) $(1, 4), (-23, -3)$

13) $(9, 8), (5, -4)$

14) $(-11, -4), (5, 8)$

15) $(-2, -6), (-2, -12)$

16) $(-1, -4), (23, 3)$

17) $(19, 3), (7, -6)$

18) $(-5, -2), (3, 4)$

19) $(2, 6), (2, -12)$

20) $(-4, -2), (8, -2)$

✍ **Find the answer for each problem.**

21) Triangle ABC is a right triangle on the coordinate system and its vertices are $(-2, 5)$, $(-2, 1)$, and $(1, 1)$. What is the area of triangle ABC? _____

22) Three vertices of a triangle on a coordinate system are $(3, -6)$, $(-5, -12)$, and $(3, -18)$. What is the perimeter of the triangle? _____

23) Four vertices of a rectangle on a coordinate system are $(-2, 2)$, $(-2, 6)$, $(4, 2)$, and $(4, 6)$. What is its perimeter? _____

Answers of Worksheets

Finding Slope

1) 1

2) −3

3) 2

4) −4

5) 6

6) −5

7) 8

8) −9

9) −7

10) 3

11) $\frac{1}{3}$

12) $-\frac{4}{5}$

13) $\frac{1}{2}$

14) −1

15) $\frac{1}{3}$

16) $\frac{1}{8}$

17) $\frac{7}{5}$

18) 2

19) −3

20) 2

21) −1

22) 2

23) 2

24) −3

25) −1

26) $\frac{5}{2}$

27) 2

28) −11

Graphing Lines Using Line Equation

1) $y = x - 2$

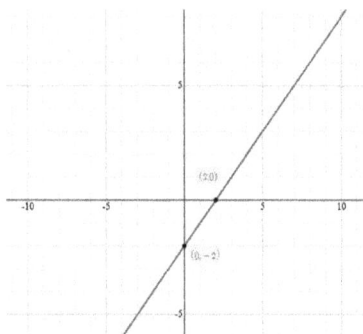

2) $y = -3x + 2$

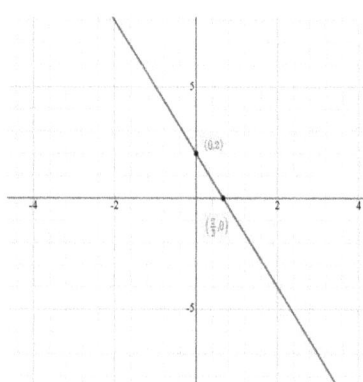

3) $x + y = 0$

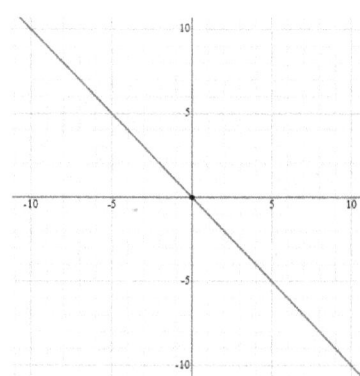

4) $x + y = -3$

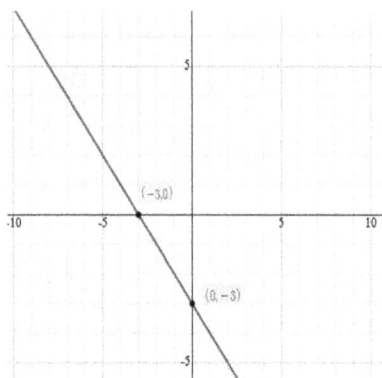

5) $2x + 3y = -4$

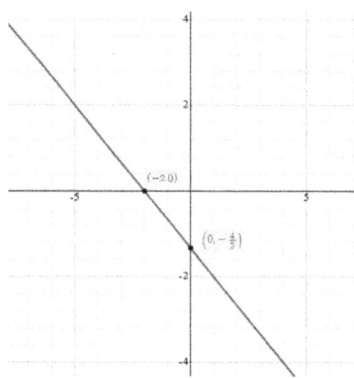

6) $y - 3x + 6 = 0$

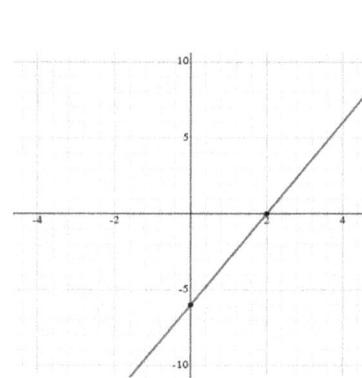

Accuplacer Subject Test – Mathematics

Writing Linear Equations

1) $y = 14x - 33$

2) $y = x + 9$

3) $y = -4x + 47$

4) $y = x - 4$

5) $y = 2x - 31$

6) $y = 9x - 43$

7) $y = -2x + 24$

8) $y = -3x + 1$

9) $y = -2x + 5$

10) $y = -2x + 6$

11) $y = 5x - 18$

12) $y = 14x - 122$

13) $y = 13x + 52$

14) $y = x + 10$

15) $y = -5x + 58$

16) $y = -5x + 5$

17) $y = 6x + 12$

18) $y = -11x - 4$

19) $y = -3x + 17$

20) $y = -5x - 11$

21) $y = -10x - 30$

22) $y = 8x + 7$

Graphing Linear Inequalities

1) $y > 4x - 5$

2) $y < 2x + 4$

3) $y \leq -5x - 2$

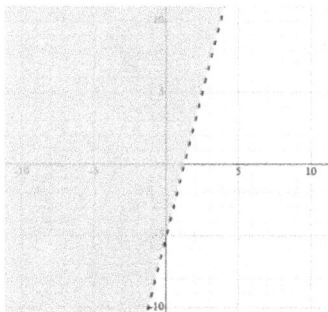

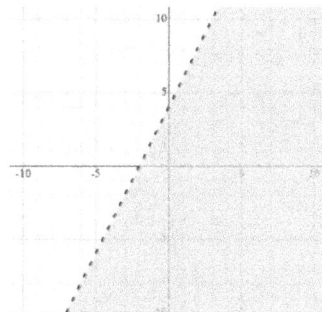

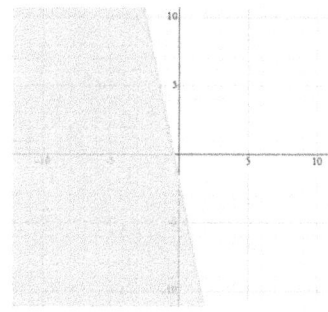

4) $4y \geq 12 + 4x$

5) $-12y < 3x - 24$

6) $5y \geq -15x + 10$

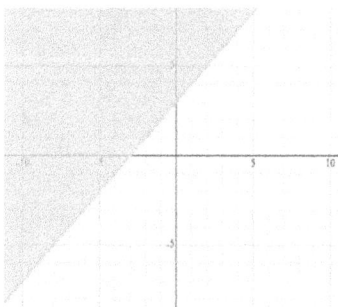

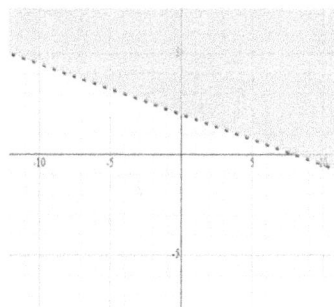

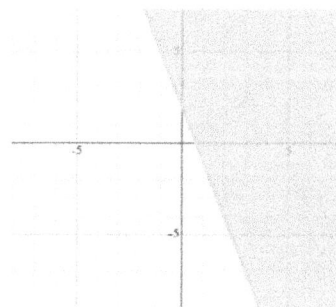

Finding Midpoint

1) $(-1, 0)$

2) $(4, 4)$

3) $(6, 4)$

4) $(-5, 1)$

5) $(8, -10)$

6) $(-3, -8)$

7) $(0, 10)$

8) $(-6, 4)$

9) $(6, -2)$

10) $(8, 6)$

11) $(-1, 6)$

12) $(4, 1)$

13) $(-4, 7.5)$

14) $(10, 2.5)$

15) $(2, 8)$

16) $(-4, 4)$

17) $(10, -10)$

18) $(4, -6)$

19) $(-1.5, -1)$

20) $(15, 5)$

21) $(4, -3)$

22) $(-9, 9)$

23) $(5, -4)$

24) $(16, 6)$

25) $(-9, 11)$

26) $(5, 3)$

27) $(14, 34)$

Finding Distance of Two Points

1) 15

2) 17

3) 2

4) 26

5) 9

6) 29

7) 25

8) 13

9) 11

10) 10

11) 10

12) 25

13) $4\sqrt{10}$

14) 20

15) 6

16) 25

17) 15

18) 10

19) 18

20) 12

21) 6 *square units*

22) 32 *units*

23) 20 *units*

Chapter 8 :

Polynomials

Topics that you'll practice in this chapter:

✓ Writing Polynomials in Standard Form

✓ Simplifying Polynomials

✓ Adding and Subtracting Polynomials

✓ Multiplying Monomials

✓ Multiplying and Dividing Monomials

✓ Multiplying a Polynomial and a Monomial

✓ Multiplying Binomials

✓ Factoring Trinomials

✓ Operations with Polynomials

Mathematics is the supreme judge; from its decisions there is no appeal.

– Tobias Dantzig

Writing Polynomials in Standard Form

✍ **Write each polynomial in standard form.**

1) $11x - 7x =$

2) $-5 + 19x - 19x =$

3) $6x^5 - 12x^3 =$

4) $12 + 17x^4 - 12 =$

5) $5x^2 + 4x - 9x^3 =$

6) $-3x^2 + 12x^5 =$

7) $5x + 8x^3 - 2x^8 =$

8) $-7x^3 + 4x - 9x^6 =$

9) $3x^2 + 22 - 6x =$

10) $3 - 4x + 9x^4 =$

11) $13x^2 + 28x - 8x^3 =$

12) $16 + 4x^2 - 2x^3 =$

13) $19x^2 - 9x + 9x^4 =$

14) $3x^4 - 7x^2 - 2x^3 =$

15) $-51 + 3x^2 - 8x^4 =$

16) $7x^2 - 8x^6 + 4x^4 - 15 =$

17) $6x^4 - 4x^5 + 16 - 3x^3 =$

18) $-2x^6 + 4x - 7x^2 - 5x =$

19) $11x^7 + 8x^5 - 5x^7 - 3x^2 =$

20) $2x^2 - 12x^5 + 8x^2 + 3x^6 =$

21) $4x^5 - 11x^7 - 6x^3 + 16x^5 =$

22) $6x^3 + 3x^5 + 34x^4 - 8x^5 =$

23) $3x(4x + 5 - 2x^2) =$

24) $12x(x^6 + 4x^3) =$

25) $5x(3x^2 + 6x + 4) =$

26) $7x(4 - 2x + 6x^5) =$

27) $3x(4x^4 - 4x^3 + 2) =$

28) $4x(2x^5 + 6x^2 - 3) =$

29) $5x(3x^4 + 4x^3 + 2x) =$

30) $2x(3x - 2x^3 + 4x^6) =$

Simplifying Polynomials

✎ **Simplify each expression.**

1) $3(4x - 20) =$

2) $5x(3x - 4) =$

3) $6x(5x - 7) =$

4) $3x(7x + 5) =$

5) $5x(4x - 3) =$

6) $6x(8x + 2) =$

7) $(3x - 2)(x - 4) =$

8) $(x - 5)(2x + 6) =$

9) $(x - 3)(x - 7) =$

10) $(3x + 4)(3x - 4) =$

11) $(5x - 4)(5x - 2) =$

12) $6x^2 + 6x^2 - 8x^4 =$

13) $3x - 2x^2 + 5x^3 + 7 =$

14) $7x + 4x^2 - 10x^3 =$

15) $12x^2 + 5x^5 - 6x^3 =$

16) $-5x^2 + 4x^6 + 6x^8 =$

17) $-12x^3 + 10x^5 - 4x^6 + 4x =$

18) $11 - 7x^2 + 4x^2 - 16x^3 + 11 =$

19) $2x^2 - 9x + 4x^3 + 15x - 10x =$

20) $13 - 7x^5 + 6x^5 - 4x^2 + 5 =$

21) $-5x^8 + x^6 - 14x^3 + 5x^8 =$

22) $(7x^4 - 4) + (7x^4 - 2x^4) =$

23) $3(3x^4 - 4x^3 - 6x^4) =$

24) $-5(x^9 + 8) - 5(10 - x^9) =$

25) $8x^3 - 9x^4 - 2x + 19 - 8x^3 =$

26) $11 - 8x^3 + 6x^3 - 7x^5 + 6 =$

27) $(5x^3 - 4x) - (6x - 2 - 6x^3) =$

28) $4x^2 - 5x^4 - x(3x^3 + 2x) =$

29) $6x + 6x^5 - 10 - 4(x^5 - 3) =$

30) $4 - 3x^4 + (6x^5 - 2x^4 + 5x^5) =$

31) $-(x^5 + 4) - 8(3 + x^5) =$

32) $(4x^3 - 3x) - (3x - 5x^3) =$

Adding and Subtracting Polynomials

✎ **Add or subtract expressions.**

1) $(-2x^2 - 3) + (3x^2 + 4) =$

2) $(4x^3 + 6) - (7 - 2x^3) =$

3) $(4x^5 + 5x^2) - (2x^5 + 15) =$

4) $(6x^3 - 2x^2) + (5x^2 - 4x) =$

5) $(10x^4 + 28x) - (34x^4 + 6) =$

6) $(7x^2 - 3) + (7x^2 + 3) =$

7) $(9x^2 + 4) - (10 - 5x^2) =$

8) $(6x^2 + x^5) - (x^5 + 4) =$

9) $(4x^3 - x) + (3x - 7x^3) =$

10) $(11x + 10) - (8x + 10) =$

11) $(15x^3 - 3x) - (3x - 4x^3) =$

12) $(4x - x^5) - (6x^5 + 8x) =$

13) $(2x^2 - 7x^7) - (4x^7 - 6x) =$

14) $(3x^2 - 5) + (8x^2 + 4x^5) =$

15) $(9x^4 + 5x^5) - (x^5 - 9x^4) =$

16) $(-4x^3 - 2x) + (9x - 5x^3) =$

17) $(4x - 3x^2) - (148x^2 + x) =$

18) $(5x - 8x^4) - (3x^4 - 4x^2) =$

19) $(8x^4 - 4) + (2x^4 - 3x^2) =$

20) $(5x^6 + 7x^3) - (x^3 - 5x^6) =$

21) $(-2x^2 + 20x^5 + 5x^4) + (12x^4 + 8x^5 + 24x^2) =$

22) $(7x^4 - 9x^7 - 6x) - (-3x^4 - 9x^7 + 6x) =$

23) $(14x + 12x^4 - 18x^6) + (20x^4 + 18x^6 - 10x) =$

24) $(5x^8 - 6x^6 - 4x) - (5x^3 + 9x^6 - 7x) =$

25) $(11x^2 - 6x^4 - 3x) - (-4x^2 - 12x^4 + 9x) =$

26) $(-5x^9 + 14x^3 + 3x^7) + (10x^7 + 26x^3 + 3x^9) =$

Multiplying Monomials

✎ **Simplify each expression.**

1) $6u^8 \times (-u^2) =$

2) $(-5p^8) \times (-2p^3) =$

3) $4xy^3z^5 \times 3z^4 =$

4) $3u^5t \times 8ut^4 =$

5) $(-5a^2) \times (-7a^3b^6) =$

6) $-3a^4b^3 \times 6a^2b =$

7) $13xy^5 \times x^4y^4 =$

8) $6p^4q^3 \times (-8pq^6) =$

9) $8s^4t^3 \times 4st^3 =$

10) $(-6x^4y^3) \times 6x^2y =$

11) $3xy^7z \times 12z^3 =$

12) $24xy \times x^2y =$

13) $13pq^4 \times (-3p^2q) =$

14) $13s^3t^4 \times st^4 =$

15) $11p^5 \times (-6p^3) =$

16) $(-8p^3q^5r) \times 3pq^4r^6 =$

17) $(-4a^4) \times (-7a^3b) =$

18) $6u^6v^2 \times (-5u^3v^4) =$

19) $9u^5 \times (-3u) =$

20) $-6xy^5 \times 4x^2y =$

21) $13y^5z^3 \times (-y^3z) =$

22) $8a^4bc^3 \times 2abc^3 =$

23) $(-7p^5q^6) \times (-5p^4q^2) =$

24) $4u^5v^3 \times (-4u^7v^3) =$

25) $17y^4z^5 \times (-y^6z) =$

26) $(-5pq^3r^2) \times 8p^2q^4r =$

27) $3ab^5c^6 \times 5a^4bc^2 =$

28) $6x^3yz^2 \times 3x^2y^7z^3 =$

Multiplying and Dividing Monomials

✎ Simplify each expression.

1) $(5x^5)(2x^2) =$

2) $(4x^4)(6x^2) =$

3) $(3x^4)(7x^4) =$

4) $(5x^6)(4x^2) =$

5) $(12x^4)(3x^6) =$

6) $(4yx^8)(8y^4x^3) =$

7) $(14x^4y)(x^3y^5) =$

8) $(-5x^3y^4)(2x^3y^5) =$

9) $(-6x^4y^2)(-3x^3y^5) =$

10) $(5x^3y)(-5x^2y^3) =$

11) $(6x^4y^3)(4x^3y^4) =$

12) $(4x^3y^2)(5x^2y^4) =$

13) $(12x^3y^6)(4x^4y^{10}) =$

14) $(15x^3y^5)(3x^4y^6) =$

15) $(7x^2y^7)(8x^6y^7) =$

16) $(-3x^3y^8)(7x^9y^4) =$

17) $\dfrac{5x^6y^6}{xy^4} =$

18) $\dfrac{19x^7y^5}{19x^6y} =$

19) $\dfrac{56x^4y^4}{8xy} =$

20) $\dfrac{81x^5y^6}{9x^4y^5} =$

21) $\dfrac{36x^7y^6}{9x^2y^3} =$

22) $\dfrac{48x^9y^7}{4x^4y^6} =$

23) $\dfrac{88x^{18}y^{12}}{11x^8y^9} =$

24) $\dfrac{30x^7y^6}{6x^8y^3} =$

25) $\dfrac{150x^7y^6}{30x^4y^6} =$

26) $\dfrac{-42x^{18}y^{14}}{6x^4y^9} =$

27) $\dfrac{-36x^7y^8}{9x^5y^8} =$

Multiplying a Polynomial and a Monomial

✎ **Find each product.**

1) $x(2x + 4) =$

2) $6(4 - 2x) =$

3) $5x(4x + 2) =$

4) $x(-4x + 5) =$

5) $8x(2x - 2) =$

6) $6(2x - 4y) =$

7) $7x(5x - 5) =$

8) $3x(12x + 2y) =$

9) $4x(x + 6y) =$

10) $11x(3x + 4y) =$

11) $7x(3x + 2) =$

12) $10x(4x - 10y) =$

13) $9x(3x - 2y) =$

14) $7x(x - 4y + 6) =$

15) $8x(2x^2 + 5y^2) =$

16) $12x(2x + 3y) =$

17) $4(2x^4 - 4y^4) =$

18) $4x(-3x^2y + 4y) =$

19) $-4(5x^3 - 2xy + 4) =$

20) $4(x^2 - 5xy - 6) =$

21) $8x(2x^3 - 5xy + 2x) =$

22) $-6x(-2x^3 - 6x + 2xy) =$

23) $3(2x^2 + xy - 9y^2) =$

24) $4x(5x^3 - 3x + 7) =$

25) $6(3x^{22} - 2x - 5) =$

26) $x^2(-2x^3 + 4x + 3) =$

27) $x^2(4x^3 + 10 - 2x) =$

28) $4x^4(3x^3 - 2x + 5) =$

29) $2x^2(4x^4 - 5xy + 7y^3) =$

30) $5x^2(5x^4 - 3x + 9) =$

31) $7x^2(6x^2 + 3x - 6) =$

32) $4x(x^3 - 4xy + 2y^2) =$

Multiplying Binomials

✍ **Find each product.**

1) $(x + 3)(x + 6) =$

2) $(x - 4)(x + 3) =$

3) $(x - 3)(x - 8) =$

4) $(x + 8)(x + 9) =$

5) $(x - 2)(x - 12) =$

6) $(x + 5)(x + 5) =$

7) $(x - 6)(x + 7) =$

8) $(x - 8)(x - 3) =$

9) $(x + 7)(x + 12) =$

10) $(x - 4)(x + 8) =$

11) $(x + 8)(x + 8) =$

12) $(x + 2)(x + 7) =$

13) $(x - 6)(x + 6) =$

14) $(x - 5)(x + 5) =$

15) $(x + 11)(x + 11) =$

16) $(x + 6)(x + 9) =$

17) $(x - 2)(x + 2) =$

18) $(x - 4)(x + 7) =$

19) $(3x + 5)(x + 6) =$

20) $(5x - 6)(4x + 8) =$

21) $(x - 7)(3x + 7) =$

22) $(x - 9)(x - 4) =$

23) $(x - 12)(x + 2) =$

24) $(2x - 4)(5x + 4) =$

25) $(3x - 8)(x + 8) =$

26) $(7x - 2)(6x + 3) =$

27) $(4x + 5)(3x + 5) =$

28) $(7x - 4)(9x + 4) =$

29) $(x + 2)(2x - 8) =$

30) $(5x - 4)(5x + 4) =$

31) $(3x + 2)(3x - 7) =$

32) $(x^2 + 8)(x^2 - 8) =$

Factoring Trinomials

✎ **Factor each trinomial.**

1) $x^2 + 8x + 12 =$

2) $x^2 - 6x + 5 =$

3) $x^2 + 15x + 36 =$

4) $x^2 - 12x + 35 =$

5) $x^2 - 11x + 18 =$

6) $x^2 - 9x + 18 =$

7) $x^2 + 18x + 72 =$

8) $x^2 - x - 72 =$

9) $x^2 + 4x - 21 =$

10) $x^2 - 13x + 22 =$

11) $x^2 + 2x - 24 =$

12) $x^2 - 3x - 40 =$

13) $x^2 - 3x - 70 =$

14) $x^2 + 26x + 169 =$

15) $4x^2 - 7x - 15 =$

16) $x^2 - 14x + 33 =$

17) $10x^2 + 5x - 15 =$

18) $6x^2 - 4x - 42 =$

19) $x^2 + 12x + 36 =$

20) $5x^2 + 17x - 12 =$

✎ **Calculate each problem.**

21) The area of a rectangle is $x^2 - x - 56$. If the width of rectangle is $x + 7$, what is its length? _____

22) The area of a parallelogram is $4x^2 + 17x - 15$ and its height is $x + 5$. What is the base of the parallelogram? _____

23) The area of a rectangle is $6x^2 - 22x + 12$. If the width of the rectangle is $3x - 2$, what is its length? _____

Operations with Polynomials

✏ **Find each product.**

1) $4(5x + 3) = $ _____

2) $8(2x + 6) = $ _____

3) $2(5x - 2) = $ _____

4) $-4(7x - 3) = $ _____

5) $3x^2(9x + 1) = $ _____

6) $4x^6(7x - 9) = $ _____

7) $3x^4(-7x + 3) = $ _____

8) $-8x^4(5x - 8) = $ _____

9) $7(x^2 + 5x - 3) = $ _____

10) $9(5x^2 - 7x + 5) = $ _____

11) $3(3x^2 + 3x + 2) = $ _____

12) $5x(3x^2 + 5x + 8) = $ _____

13) $(5x + 7)(3x - 3) = $ _____

14) $(9x + 3)(3x - 5) = $ _____

15) $(6x + 3)(4x - 2) = $ _____

16) $(7x - 2)(3x + 5) = $ _____

✏ **Calculate each problem.**

17) The measures of two sides of a triangle are $(2x + 5y)$ and $(6x - 3y)$. If the perimeter of the triangle is $(13x + 4y)$, what is the measure of the third side? _____

18) The height of a triangle is $(8x + 5)$ and its base is $(4x - 3)$. What is the area of the triangle? _____

19) One side of a square is $(6x + 2)$. What is the area of the square? _____

20) The length of a rectangle is $(5x - 8y)$ and its width is $(15x + 8y)$. What is the perimeter of the rectangle? _____

21) The side of a cube measures $(x + 2)$. What is the volume of the cube? _____

22) If the perimeter of a rectangle is $(28x + 6y)$ and its width is $(5x + 2y)$, what is the length of the rectangle? _____

Answers of Worksheets

Writing Polynomials in Standard Form

1) $4x$

2) -5

3) $6x^5 - 12x^3$

4) $14x^4$

5) $-9x^3 + 5x^2 + 4x$

6) $12x^5 - 3x^2$

7) $-2x^8 + 8x^3 + 5x$

8) $-9x^6 - 7x^3 + 4x$

9) $3x^2 - 6x + 22$

10) $9x^4 - 4x + 3$

11) $-8x^3 + 13x^2 + 28x$

12) $-2x^3 + 4x^2 + 16$

13) $9x^4 + 19x^2 - 9x$

14) $3x^4 - 2x^3 - 7x^2$

15) $-8x^4 + 3x^2 - 51$

16) $-8x^6 + 4x^4 + 7x^2 - 15$

17) $-4x^5 + 6x^4 - 3x^3 + 16$

18) $-2x^6 - 7x^2 - x$

19) $6x^7 + 8x^5 - 3x^2$

20) $3x^6 - 12x^5 + 10x^2$

21) $-11x^7 + 20x^5 - 6x^3$

22) $-5x^5 + 34x^4 + 6x^3$

23) $-6x^3 + 12x^2 + 15x$

24) $12x^7 + 48x^4$

25) $15x^3 + 30x^2 + 20x$

26) $42x^6 - 14x^2 + 28x$

27) $12x^5 - 12x^4 + 6x$

28) $8x^6 + 24x^3 - 12x$

29) $15x^5 + 20x^4 + 10x^2$

30) $8x^7 - 4x^4 + 6x^2$

Simplifying Polynomials

1) $12x - 60$

2) $15x^2 - 20x$

3) $30x^2 - 42x$

4) $21x^2 + 15x$

5) $20x^2 - 15x$

6) $48x^2 + 12x$

7) $3x^2 - 14x + 8$

8) $2x^2 - 4x - 30$

9) $x^2 - 10x + 21$

10) $9x^2 - 16$

11) $25x^2 - 30x + 8$

12) $-8x^4 + 12x^2$

13) $5x^3 - 2x^2 + 3x + 7$

14) $-10x^3 + 4x^2 + 7x$

15) $5x^5 - 6x^3 + 12x^2$

16) $6x^8 + 4x^6 - 5x^2$

17) $-4x^6 + 10x^5 - 12x^3 + 4x$

18) $-16x^3 - 3x^2 + 22$

19) $4x^3 + 2x^2 - 4x$

20) $-x^5 - 4x^2 + 18$

21) $x^6 - 14x^3$

22) $12x^4 - 4$

23) $-9x^4 - 12x^3$

24) -90

25) $-9x^4 - 2x + 19$

26) $-7x^5 - 2x^3 + 17$

27) $11x^3 - 10x + 2$

28) $-8x^4 + 2x^2$

29) $2x^5 + 6x + 2$

30) $11x^5 - 5x^4 + 4$

31) $-9x^5 - 28$

32) $9x^3 - 6x$

Adding and Subtracting Polynomials

1) $x^2 + 1$

2) $6x^3 - 1$

3) $2x^5 + 5x^2 - 15$

4) $6x^3 + 3x^2 - 4x$

5) $-24x^4 + 28x - 6$

6) $14x^2$

7) $14x^2 - 6$

8) $6x^2 - 4$

9) $-3x^3 + 2x$

10) $3x$

11) $19x^3 - 6x$

12) $-7x^5 - 4x$

13) $-11x^7 + 2x^2 + 6x$

14) $4x^5 + 11x^2 - 5$

15) $4x^5 + 18x^4$

16) $-9x^3 + 7x$

17) $-151x^2 + 3x$

18) $-11x^4 + 4x^2 + 5x$

19) $10x^4 - 3x^2 - 4$

20) $10x^6 + 6x^3$

21) $28x^5 + 17x^4 + 22x^2$

22) $10x^4 - 12x$

23) $32x^4 + 4x$

24) $5x^8 - 15x^6 - 5x^3 + 3x$

25) $6x^4 + 15x^2 - 12x$

26) $-2x^9 + 13x^7 + 40x^3$

Multiplying Monomials

1) $-6u^{10}$

2) $10p^{11}$

3) $12xy^3z^9$

4) $24u^6t^5$

5) $35a^5b^6$

6) $-18a^6b^4$

7) $13x^5y^9$

8) $-48p^5q^9$

9) $32s^5t^6$

10) $-36x^6y^4$

11) $36xy^7z^4$

12) $24px^3y^2$

13) $-39p^3q^5$

14) $13s^4t^8$

15) $-66p^8$

16) $-24p^4q^9r^7$

17) $28a^7b$

18) $-30u^9v^6$

19) $-27u^6$

20) $-24x^3y^6$

21) $-13y^8z^4$

22) $16a^5b^2c^6$

23) $35p^9q^8$

24) $-16u^{12}v^6$

25) $-17y^{10}z^6$

26) $-40p^3q^7r^3$

27) $15a^5b^6c^8$

28) $18x^5y^8z^5$

Multiplying and Dividing Monomials

1) $10x^7$

2) $24x^6$

3) $21x^8$

4) $20x^8$

5) $36x^{10}$

6) $32x^{11}y^5$

7) $14x^7y^6$

8) $-10x^6y^9$

9) $18x^7y^7$

10) $-25x^5y^4$

11) $24x^7y^7$

12) $20x^5y^6$

13) $48x^7y^{16}$ 18) xy^4 23) $8x^{10}y^3$

14) $45x^7y^{11}$ 19) $7x^3y^3$ 24) $5x^{-1}y^3$

15) $56x^8y^{14}$ 20) $9xy$ 25) $5x^3$

16) $-21x^{12}y^{12}$ 21) $4x^5y^3$ 26) $-7x^{14}y^5$

17) $5x^5y^2$ 22) $12x^5y$ 27) $-4x^2$

Multiplying a Polynomial and a Monomial

1) $2x^2 + 4x$ 17) $8x^4 - 16y^4$

2) $-12x + 24$ 18) $-12x^3y + 16xy$

3) $20x^2 + 10x$ 19) $-20x^3 + 8xy - 16$

4) $-4x^2 + 5x$ 20) $4x^2 - 20xy - 24$

5) $16x^2 - 16x$ 21) $16x^4 - 40x^2y + 16x^2$

6) $12x - 24y$ 22) $12x^4 + 36x^2 - 12x^2y$

7) $35x^2 - 35x$ 23) $6x^2 + 3xy - 27y^2$

8) $36x^2 + 6xy$ 24) $20x^4 - 12x^2 + 28x$

9) $4x^2 + 24xy$ 25) $18x^{22} - 12x - 30$

10) $33x^2 + 44xy$ 26) $-2x^5 + 4x^3 + 3x^2$

11) $21x^2 + 14x$ 27) $4x^5 - 2x^3 + 10x^2$

12) $40x^2 - 100xy$ 28) $12x^7 - 8x^5 + 20x^4$

13) $27x^2 - 18xy$ 29) $8x^6 - 10x^3y + 14x^2y^3$

14) $7x^2 - 28xy + 42x$ 30) $25x^6 - 15x^3 + 45x^2$

15) $16x^3 + 40xy^2$ 31) $42x^4 + 21x^3 - 42x^2$

16) $24x^2 + 36xy$ 32) $4x^4 - 16x^2y + 8xy^2$

Multiplying Binomials

1) $x^2 + 9x + 18$ 8) $x^2 - 11x + 24$

2) $x^2 - x - 12$ 9) $x^2 + 19x + 84$

3) $x^2 - 11x + 24$ 10) $x^2 + 4x - 32$

4) $x^2 + 17x + 72$ 11) $x^2 + 16x + 64$

5) $x^2 - 14x + 24$ 12) $x^2 + 9x + 14$

6) $x^2 + 10x + 25$ 13) $x^2 - 36$

7) $x^2 + x - 42$ 14) $x^2 - 25$

15) $x^2 + 22x + 121$

16) $x^2 + 15x + 54$

17) $x^2 - 4$

18) $x^2 + 3x - 28$

19) $3x^2 + 23x + 30$

20) $20x^2 + 16x - 48$

21) $3x^2 - 14x - 49$

22) $x^2 - 13x + 36$

23) $x^2 - 10x - 24$

24) $10x^2 - 12x - 16$

25) $3x^2 + 16x - 64$

26) $42x^2 + 9x - 6$

27) $12x^2 + 35x + 25$

28) $63x^2 - 8x - 16$

29) $2x^2 - 4x - 16$

30) $25x^2 - 16$

31) $9x^2 - 15x - 14$

32) $x^4 - 64$

Factoring Trinomials

1) $(x + 6)(x + 2)$

2) $(x - 5)(x - 1)$

3) $(x + 12)(x + 3)$

4) $(x - 5)(x - 7)$

5) $(x - 2)(x - 9)$

6) $(x - 6)(x - 3)$

7) $(x + 6)(x + 12)$

8) $(x + 8)(x - 9)$

9) $(x - 3)(x + 7)$

10) $(x - 11)(x - 2)$

11) $(x - 4)(x + 6)$

12) $(x - 8)(x + 5)$

13) $(x + 7)(x - 10)$

14) $(x + 13)(x + 13)$

15) $(4x + 5)(x - 3)$

16) $(x - 11)(x - 3)$

17) $(5x - 5)(2x + 3)$

18) $(2x - 6)(3x + 7)$

19) $(x + 6)(x + 6)$

20) $(5x - 3)(x + 4)$

21) $(x - 8)$

22) $(4x - 3)$

23) $(2x - 6)$

Operations with Polynomials

1) $20x + 12$

2) $16x + 48$

3) $10x - 4$

4) $-28x + 12$

5) $27x^3 + 3x^2$

6) $28x^7 - 36x^6$

7) $-21x^5 + 9x^4$

8) $-40x^5 + 64x^4$

9) $7x^2 + 35x - 21$

10) $45x^2 - 63x + 45$

11) $9x^2 + 9x + 6$

12) $15x^3 + 25x^2 + 40x$

13) $15x^2 + 6x - 21$

14) $27x^2 - 36x - 15$

15) $24x^2 - 6$

16) $21x^2 + 29x - 10$

17) $(5x + 2y)$

18) $16x^2 - 2x - \frac{15}{2}$

19) $36x^2 + 24x + 4$

20) $40x$

21) $x^3 + 6x^2 + 12x + 8$

22) $(9x + y)$

Chapter 9 :

Complex Numbers

Topics that you'll practice in this chapter:

✓ Adding and Subtracting Complex Numbers

✓ Multiplying and Dividing Complex Numbers

✓ Graphing Complex Numbers

✓ Rationalizing Imaginary Denominators

Mathematics is a hard thing to love. It has the unfortunate habit, like a rude dog, of turning its most unfavorable side towards you when you first make contact with it. — *David Whiteland*

Adding and Subtracting Complex Numbers

✎ **Simplify.**

1) $(7i) - (3i) =$

2) $(5i) + (4i) =$

3) $(2i) + (8i) =$

4) $(-8i) - (3i) =$

5) $(14i) + (6i) =$

6) $(6i) - (-10i) =$

7) $(-2i) + (-5i) =$

8) $(13i) - (5i) =$

9) $(-22i) - (11i) =$

10) $(-4i) + (2 + 6i) =$

11) $(10 - 5i) + (-3i) =$

12) $(-8i) + (6 + 12i) =$

13) $1 + (5 - 4i) =$

14) $(13i) - (-8 + 2i) =$

15) $(8 + 12i) - (-10i) =$

16) $(10 + i) + (-5i) =$

17) $(11i) - (-7i + 9) =$

18) $(10i + 12) + (-2i) =$

19) $(20) - (16 + 4i) =$

20) $(3 + 3i) + (8 + 4i) =$

21) $(15 - 7i) + (3 + 4i) =$

22) $(12 + 6i) + (10 + 17i) =$

23) $(-5 + 6i) - (-16 - 12i) =$

24) $(-4 + 14i) - (-9 + 11i) =$

25) $(-22 + 4i) - (- 7 - 22i) =$

26) $(-26 - 18i) + (3 + 34i) =$

27) $(-19 - 13i) - (-7 - 20i) =$

28) $-21 + (5i) + (-32 + 14i) =$

29) $30 - (7i) + (3 - 11i) =$

30) $28 + (-32 - 10i) - 7 =$

31) $(-44i) + (2 - 7i) + 9 =$

32) $(-21i) - (12 - 9i) + 21i =$

Multiplying and Dividing Complex Numbers

✎ **Simplify.**

1) $(5i)(-3i) =$

2) $(-8i)(2i) =$

3) $(3i)(-3i)(-3i) =$

4) $(6i)(-6i) =$

5) $(-3 - 4i)(2 + i) =$

6) $(5 - 2i)^2 =$

7) $(5 - 2i)(6 - 4i) =$

8) $(1 + 6i)^2 =$

9) $(5i)(-3i)(2 - 4i) =$

10) $(11 - 2i)(2 - 4i) =$

11) $(-3 + i)(6 + 5i) =$

12) $(2 - 8i)(6 - 4i) =$

13) $3(4i) - (6i)(-2 + 5i) =$

14) $\dfrac{5}{-25i} =$

15) $\dfrac{3-4i}{-5i} =$

16) $\dfrac{6+12i}{2i} =$

17) $\dfrac{20i}{-5+4i} =$

18) $\dfrac{-6-9i}{4i} =$

19) $\dfrac{4i}{8-2i} =$

20) $\dfrac{4-7i}{6-2i} =$

21) $\dfrac{3-2i}{-1-1i} =$

22) $\dfrac{-5-5i}{-4-i} =$

23) $\dfrac{-6+2i}{-10-4i} =$

24) $\dfrac{-8-4i}{-2+4i} =$

25) $\dfrac{2+3i}{1-4i} =$

Graphing Complex Numbers

✍ **Identify each complex number graphed.**

1)

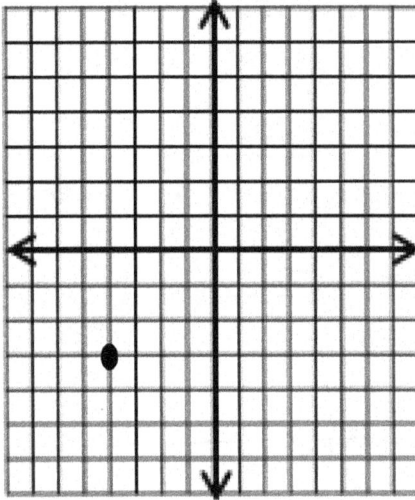

2)

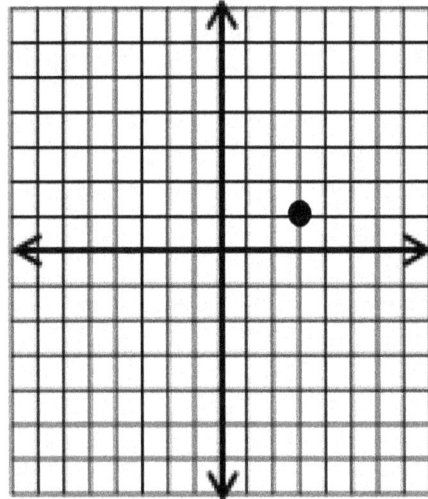

3)

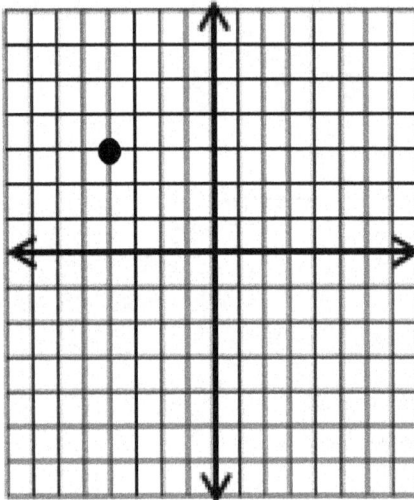

4)

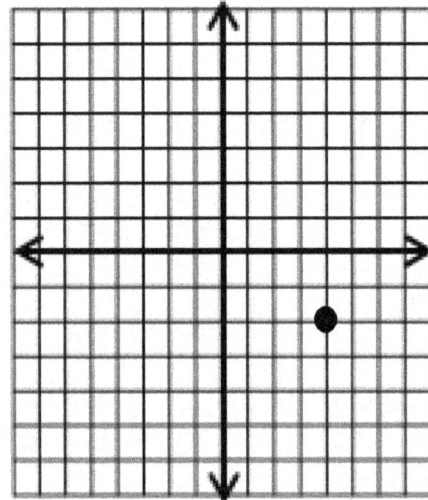

Rationalizing Imaginary Denominators

✎ **Simplify.**

1) $\dfrac{-7}{-7i} =$

2) $\dfrac{-3}{-15i} =$

3) $\dfrac{-3}{-39i} =$

4) $\dfrac{24}{-3i} =$

5) $\dfrac{5}{2i} =$

6) $\dfrac{16}{-4i} =$

7) $\dfrac{14}{-6i} =$

8) $\dfrac{-17}{3i} =$

9) $\dfrac{4x}{5yi} =$

10) $\dfrac{10-10i}{-2i} =$

11) $\dfrac{5-11i}{-i} =$

12) $\dfrac{21+4i}{4i} =$

13) $\dfrac{8i}{-1+4i} =$

14) $\dfrac{10i}{-6+8i} =$

15) $\dfrac{-25-5i}{-5+5i} =$

16) $\dfrac{-7-2i}{3+1i} =$

17) $\dfrac{-12-6i}{8-6i} =$

18) $\dfrac{-14+7i}{-7i} =$

19) $\dfrac{12+3i}{3i} =$

20) $\dfrac{-2-i}{4-3i} =$

21) $\dfrac{-11+4i}{-5i} =$

22) $\dfrac{8+2i}{-5-2i} =$

23) $\dfrac{-9-5i}{-8-2i} =$

24) $\dfrac{4i-1}{-5-2i} =$

Answers of Worksheets – Chapter 9

Adding and Subtracting Complex Numbers

1) $4i$	9) $-33i$	17) $-9 + 18i$	25) $-15 + 26i$
2) $9i$	10) $2 + 2i$	18) $12 + 8i$	26) $-23 + 16i$
3) $10i$	11) $10 - 8i$	19) $4 - 4i$	27) $-12 + 7i$
4) $-11i$	12) $6 + 4i$	20) $11 + 7i$	28) $-53 + 19i$
5) $20i$	13) $6 - 4i$	21) $18 - 3i$	29) $33 - 18i$
6) $16i$	14) $8 + 11i$	22) $22 + 23i$	30) $-11 - 10i$
7) $-7i$	15) $8 + 22i$	23) $11 + 18i$	31) $11 - 51i$
8) $8i$	16) $10 - 4i$	24) $5 + 3i$	32) $-12 + 9i$

Multiplying and Dividing Complex Numbers

1) 15	9) $30 - 60i$	16) $-6 + 3i$	22) $\frac{25}{17} + \frac{15}{17}i$
2) 16	10) $14 - 48i$	17) $\frac{80}{41} - \frac{100}{41}i$	23) $\frac{13}{29} - \frac{11}{29}i$
3) $-27i$	11) $-23 - 9i$	18) $\frac{9}{4} - \frac{3}{2}i$	24) $2i$
4) 36	12) $-20 - 56i$	19) $-\frac{2}{17} + \frac{8}{17}i$	25) $-\frac{10}{17} + \frac{11}{17}i$
5) $-2 - 11i$	13) $30 + 24i$	20) $\frac{19}{20} - \frac{17}{20}i$	
6) $21 - 20i$	14) $\frac{i}{5}$	21) $-\frac{1}{2} + \frac{5}{2}i$	
7) $22 - 32i$	15) $\frac{4}{5} + \frac{3}{5}i$		
8) $-35 + 12i$			

Graphing Complex Numbers

1) $-4 - 3i$	2) $3 + i$	3) $-4 + 3i$	4) $4 - 2i$

Rationalizing Imaginary Denominators

1) $-i$	7) $\frac{7}{3}i$	13) $\frac{32}{17} - \frac{8}{17}i$	19) $1 - 4i$
2) $-\frac{1}{5}i$	8) $\frac{17}{3}i$	14) $\frac{4}{5} - \frac{3}{5}i$	20) $-\frac{1}{5} - \frac{2}{5}i$
3) $\frac{-1}{13}i$	9) $-\frac{4x}{5y}i$	15) $2 + 3i$	21) $-\frac{4}{5} - \frac{11}{5}i$
4) $8i$	10) $5 + 5i$	16) $-\frac{23}{10} + \frac{1}{10}i$	22) $-\frac{44}{29} + \frac{6}{29}i$
5) $-\frac{5}{2}i$	11) $11 + 5i$	17) $-\frac{3}{5} - \frac{6}{5}i$	23) $\frac{41}{34} + \frac{11}{34}i$
6) $4i$	12) $1 - \frac{21}{4}i$	18) $-1 - 2i$	24) $-\frac{3}{29} - \frac{22}{29}i$

Chapter 10 :

Functions Operations and Quadratic

Topics that you'll practice in this chapter:

- ✓ Evaluating Function
- ✓ Adding and Subtracting Functions
- ✓ Multiplying and Dividing Functions
- ✓ Composition of Functions
- ✓ Quadratic Equation
- ✓ Solving Quadratic Equations
- ✓ Quadratic Formula and the Discriminant
- ✓ Quadratic Inequalities
- ✓ Graphing Quadratic Functions
- ✓ Domain and Range of Radical Functions
- ✓ Solving Radical Equations

It's fine to work on any problem, so long as it generates interesting mathematics along the way –
even if you don't solve it at the end of the day." – Andrew Wiles

Evaluating Function

📖 **Write each of following in function notation.**

1) $h = -8x + 3$

2) $k = 2a - 14$

3) $d = 11t$

4) $y = \frac{5}{12}x - \frac{7}{12}$

5) $m = 24n - 210$

6) $c = p^2 - 5p + 10$

📖 **Evaluate each function.**

7) $f(x) = 2x - 7$, find $f(-3)$

8) $g(x) = \frac{1}{9}x + 12$, find $f(18)$

9) $h(x) = -4x + 9$, find $f(3)$

10) $f(x) = -x + 19$, find $f(-3)$

11) $f(a) = 7a - 12$, find $f(3)$

12) $h(x) = 14 - 3x$, find $f(-4)$

13) $g(n) = 6n - 10$, find $f(2)$

14) $f(x) = -11x - 4$, find $f(-1)$

15) $k(n) = -20 - 3.5n$, find $f(2)$

16) $f(x) = -0.7x + 3.3$, find $f(-7)$

17) $g(n) = \frac{11n+8}{n}$, find $g(2)$

18) $g(n) = \sqrt{3n} + 12$, find $g(3)$

19) $h(x) = x^{-2} - 7$, find $h(\frac{1}{9})$

20) $h(n) = n^{-3} + 11$, find $h(\frac{1}{4})$

21) $h(n) = n^3 - 2$, find $h(\frac{1}{2})$

22) $h(n) = n^2 - 4$, find $h(-\frac{1}{3})$

23) $h(n) = 4n^2 - 13$, find $h(-5)$

24) $h(n) = -2n^3 - 6n$, find $h(2)$

25) $g(n) = \sqrt{16n^2} - \sqrt{n}$, find $g(4)$

26) $h(a) = \frac{-14a+9}{3a}$, find $h(-b)$

27) $k(a) = 12a - 14$, find $k(a - 3)$

28) $h(x) = \frac{1}{9}x + 18$, find $h(-18x)$

29) $h(x) = 8x^2 + 16$, find $h(\frac{x}{2})$

30) $h(x) = x^4 - 20$, find $h(-2x)$

Adding and Subtracting Functions

✎**Perform the indicated operation.**

1) $f(x) = 2x + 3$

   $g(x) = x + 7$

   Find $(f - g)(2)$

2) $g(a) = -5a - 8$

   $f(a) = -3a - 5$

   Find $(g - f)(-2)$

3) $h(t) = 4t + 3$

   $g(t) = 4t + 7$

   Find $(h - g)(t)$

4) $g(a) = -6a - 10$

   $f(a) = 3a^2 + 9$

   Find $(g - f)(x)$

5) $g(x) = \frac{5}{6}x - 23$

   $h(x) = \frac{5}{12}x + 25$

   Find $g(12) - h(12)$

6) $h(x) = \sqrt{3x} - 2$

   $g(x) = \sqrt{3x} + 5$

   Find $(h + g)(12)$

7) $f(x) = x^{-1}$

   $g(x) = x^2 + \frac{5}{x}$

   Find $(f - g)(-3)$

8) $h(n) = n^2 + 2$

   $g(n) = -4n + 6$

   Find $(h - g)(2a)$

9) $g(x) = -2x^2 - 5 - 4x$

   $f(x) = 7 + 2x$

   Find $(g - f)(3x)$

10) $g(t) = 11t - 4$

    $f(t) = -2t^2 + 5$

    Find $(g + f)(-t)$

11) $f(x) = 8x + 9$

    $g(x) = -5x^2 + 3x$

    Find $(f - g)(-x^2)$

12) $f(x) = -3x^4 - 5x$

    $g(x) = 2x^4 + 5x + 22$

    Find $(f + g)(3x^2)$

Multiplying and Dividing Functions

✎ **Perform the indicated operation.**

1) $g(x) = -2x - 1$

   $f(x) = 4x + 3$

   Find $(g.f)(2)$

2) $f(x) = 5x$

   $h(x) = -2x + 3$

   Find $(f.h)(-2)$

3) $g(a) = 5a - 2$

   $h(a) = 2a - 3$

   Find $(g.h)(-3)$

4) $f(x) = 2x - 7$

   $h(x) = x - 5$

   Find $\left(\frac{f}{h}\right)(4)$

5) $f(x) = 8a^2$

   $g(x) = 3 + 2a$

   Find $\left(\frac{f}{g}\right)(2)$

6) $g(a) = \sqrt{4a} + 2$

   $f(a) = (-a)^4 + 1$

   Find $\left(\frac{g}{f}\right)(1)$

7) $g(t) = t^3 + 1$

   $h(t) = 5t - 2$

   Find $(g.h)(-2)$

8) $g(n) = n^2 + 2n - 4$

   $h(n) = -5n + 3$

   Find $(g.h)(1)$

9) $g(a) = (a - 3)^2$

   $f(a) = a^2 + 4$

   Find $\left(\frac{g}{f}\right)(3)$

10) $g(x) = -3x^2 + \frac{4}{5}x + 9$

    $f(x) = x^2 - 24$

    Find $\left(\frac{g}{f}\right)(5)$

11) $f(x) = 2x^3 - 5x^2 + 1$

    $g(x) = 3x - 1$

    Find $(f.g)(x)$

12) $f(x) = 5x - 2$

    $g(x) = x^3 - 2x$

    Find $(f.g)(x^2)$

Composition of Functions

Using $f(x) = 2x - 5$ and $g(x) = -2x$, find:

1) $f(g(2)) =$

2) $f(g(-1)) =$

3) $g(f(-4)) =$

4) $g(f(5)) =$

5) $f(g(3)) =$

6) $g(f(0)) =$

Using $f(x) = -\frac{1}{4}x + \frac{3}{4}$ and $g(x) = 2x^2$, find:

7) $g(f(-2)) =$

8) $g(f(4)) =$

9) $g(g(1)) =$

10) $f(f(1)) =$

11) $g(f(-4)) =$

12) $f(g(x)) =$

Using $f(x) = -2x + 2$ and $g(x) = x + 1$, find:

13) $g(f(1)) =$

14) $f(f(0)) =$

15) $f(g(-1)) =$

16) $f(g(-3)) =$

17) $g(f(2)) =$

18) $f(g(x)) =$

Using $f(x) = \sqrt{x + 9}$ and $g(x) = x - 9$, find:

19) $f(g(9)) =$

20) $g(f(-9)) =$

21) $f(g(4)) =$

22) $f(f(7)) =$

23) $g(f(-5)) =$

24) $g(g(0)) =$

Quadratic Equation

✎ Multiply.

1) $(x - 4)(x + 6) =$ _____

2) $(x + 5)(x + 7) =$ _____

3) $(x - 6)(x + 8) =$ _____

4) $(x + 2)(x - 9) =$ _____

5) $(x - 7)(x - 8) =$ _____

6) $(3x + 2)(x - 3) =$ _____

7) $(4x - 3)(x + 2) =$ _____

8) $(4x - 5)(x + 1) =$ _____

9) $(7x + 1)(x - 6) =$ _____

10) $(5x + 1)(3x - 3) =$ _____

✎ Factor each expression.

11) $x^2 - 2x - 8 =$ _____

12) $x^2 + 8x + 15 =$ _____

13) $x^2 - 2x - 24 =$ _____

14) $x^2 - 10x + 21 =$ _____

15) $x^2 + 10x + 21 =$ _____

16) $4x^2 + 9x + 5 =$ _____

17) $5x^2 + 13x - 6 =$ _____

18) $5x^2 + 17x - 12 =$ _____

19) $2x^2 + 7x + 5 =$ _____

20) $9x^2 - 21x + 6 =$ _____

✎ Calculate each equation.

21) $(x + 6)(x - 3) = 0$

22) $(x + 1)(x + 8) = 0$

23) $(3x + 6)(x + 5) = 0$

24) $(2x - 2)(4x + 8) = 0$

25) $x^2 + x + 10 = 22$

26) $x^2 + 11x + 36 = 12$

27) $2x^2 + 9x + 9 = 5$

28) $x^2 + 3x - 24 = 4$

29) $5x^2 + 5x - 40 = 20$

30) $8x^2 + 8x = 48$

Solving Quadratic Equations

✎ **Solve each equation by factoring or using the quadratic formula.**

1) $(x + 9)(x - 1) = 0$

2) $(x + 7)(x + 6) = 0$

3) $(x - 8)(x + 3) = 0$

4) $(x - 6)(x - 4) = 0$

5) $(x + 2)(x + 12) = 0$

6) $(5x + 4)(x + 7) = 0$

7) $(6x + 1)(4x + 5) = 0$

8) $(2x + 7)(x + 8) = 0$

9) $(x + 6)(3x + 15) = 0$

10) $(12x + 2)(x + 8) = 0$

11) $x^2 = 8x$

12) $x^2 - 16 = 0$

13) $3x^2 + 6 = 9x$

14) $-2x^2 - 8 = 10x$

15) $5x^2 + 40x = 45$

16) $x^2 + 10x = 24$

17) $x^2 + 6x = 16$

18) $x^2 + 9x = -18$

19) $x^2 + 13x = -36$

20) $x^2 + 3x - 15 = 5x$

21) $x^2 + 8x + 7 = -8$

22) $3x^2 - 11x = -9 + x$

23) $10x^2 + 3 = 27x - 15$

24) $7x^2 - 6x + 8 = 8$

25) $2x^2 - 12 = -3x + 2$

26) $10x^2 - 26x - 3 = -15$

27) $3x^2 + 21 = -16x + 5$

28) $x^2 + 15x - 10 = -66$

29) $3x^2 - 8x - 8 = 4 + x$

30) $2x^2 + 6x - 24 = 12$

31) $3x^2 - 33x + 54 = -18$

32) $-10x^2 - 15x - 9 = -9 - 27x^2$

Quadratic Formula and the Discriminant

✍ **Find the value of the discriminant of each quadratic equation.**

1) $3x(x - 8) = 0$

2) $2x^2 + 6x - 4 = 0$

3) $x^2 + 6x + 7 = 0$

4) $x^2 - x + 3 = 0$

5) $x^2 + 4x - 3 = 0$

6) $2x^2 + 6x - 10 = 0$

7) $3x^2 + 7x + 5 = 0$

8) $x^2 - 6x - 4 = 0$

9) $2x^2 + 8x + 3 = 0$

10) $x^2 + 7x - 5 = 0$

11) $5x^2 + 2x - 3 = 0$

12) $-3x^2 - 11x + 4 = 0$

13) $-6x^2 - 12x + 8 = 0$

14) $-x^2 - 9x - 12 = 0$

15) $7x^2 - 6x - 10 = 0$

16) $-4x^2 - 2x + 8 = 0$

17) $5x^2 + 8x - 2 = 0$

18) $6x^2 - 4x = 0$

19) $3x^2 - 5x + 2 = 0$

20) $4x^2 + 9x + 3 = 0$

✍ **Find the discriminant of each quadratic equation then state the number of real and imaginary solutions.**

21) $-4x^2 - 16 = 16x$

22) $20x^2 = 20x - 5$

23) $-11x^2 - 19x = 26$

24) $22x^2 - 4x + 1 = 18x^2$

25) $-11x^2 = -15x + 8$

26) $3x^2 + 6x + 9 = 6$

27) $13x^2 - 5x - 12 = -26$

28) $-8x^2 - 32x - 25 = 7$

Graphing Quadratic Functions

✎ketch the graph of each function. Identify the vertex and axis of symmetry.

1) $y = (x + 3)^2 + 2$

2) $y = (x - 3)^2 - 2$

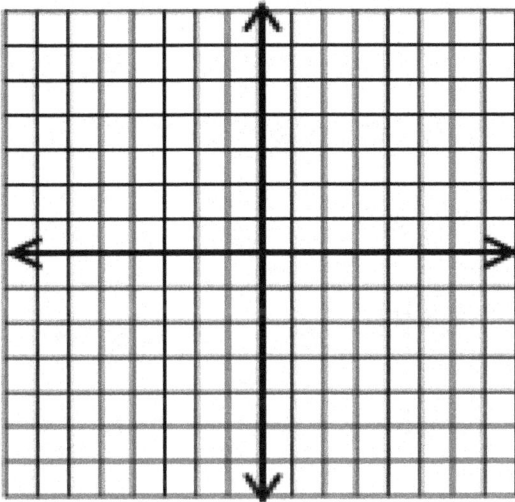

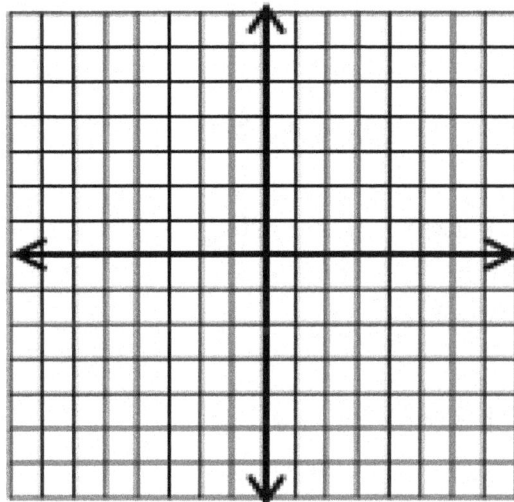

3) $y = 6 - (-x + 4)^2$

4) $y = -3x^2 - 6x + 9$

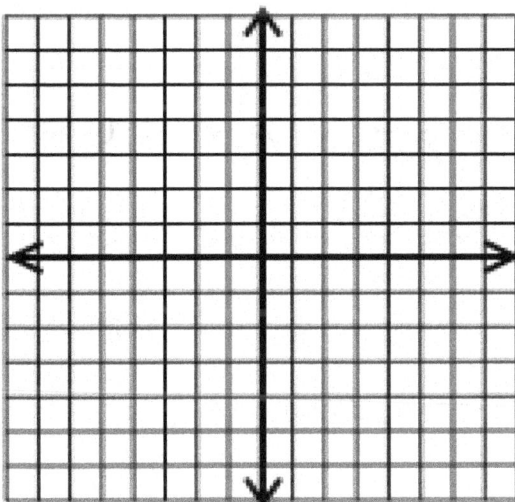

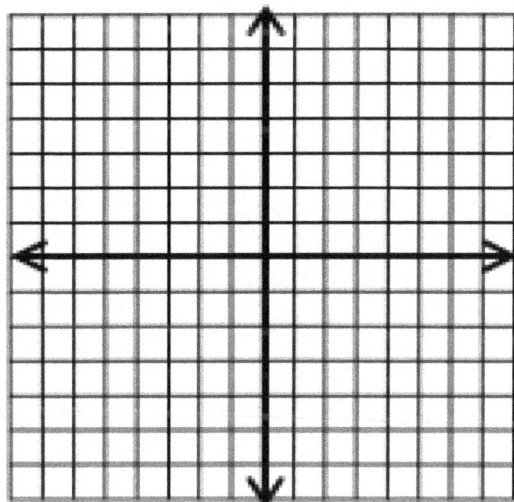

Quadratic Inequalities

✎Solve each quadratic inequality.

1) $x^2 - 25 < 0$

2) $-x^2 - 6x - 8 > 0$

3) $5x^2 + 15x + 30 < 0$

4) $x^2 + 8x + 16 > 0$

5) $2x^2 - 18x - 20 \geq 0$

6) $x^2 > -10x - 25$

7) $3x^2 + 2x + 16 \leq 0$

8) $x^2 - 5x - 14 \leq 0$

9) $x^2 - 6x - 7 \geq 0$

10) $2x^2 + 16x - 18 < 0$

11) $x^2 + 6x - 72 > 0$

12) $3x^2 - 3x - 36 > 0$

13) $x^2 - 15x + 64 \leq 0$

14) $2x^2 - 24x + 72 \leq 0$

15) $x^2 - 16x + 63 \geq 0$

16) $x^2 - 16x + 55 \geq 0$

17) $x^2 - 81 \leq 0$

18) $x^2 - 17x + 42 \geq 0$

19) $9x^2 + 14x + 36 \leq 0$

20) $4x^2 - 2x - 24 > 2x^2$

21) $5x^2 - 20x + 20 < 0$

22) $7x^2 - 6x \geq 6x^2 - 5$

23) $5x^2 - 15 > 4x^2 + 2x$

24) $3x^2 - 4x \geq 3x^2 - 9x + 15$

25) $8x^2 + 9x - 54 > 5x^2$

26) $10x^2 + 50x - 60 < 0$

27) $-x^2 + 15x - 57 \geq 0$

28) $-5x^2 + 25x + 30 \leq 0$

29) $5x^2 + 40x + 75 < 0$

30) $9x^2 + 20x + 180 \leq 0$

31) $3x^2 + 2x - 36 \geq -x$

32) $3x^2 + 9x + 9 \leq 6x^2 + 3x$

Domain and Range of Radical Functions

✍ **Identify the domain and range of each function.**

1) $y = \sqrt{x + 8} - 7$

2) $y = \sqrt[3]{3x - 5} - 4$

3) $y = \sqrt{3x - 9} + 3$

4) $y = \sqrt[3]{(4x + 6)} - 2$

5) $y = 3\sqrt{4x + 20} + 6$

6) $y = \sqrt[3]{(5x - 2)} - 11$

7) $y = 4\sqrt{9x^2 + 8} + 3$

8) $y = \sqrt[3]{(7x^2 - 2)} - 6$

9) $y = 2\sqrt{2x^3 + 16} - 3$

10) $y = \sqrt[3]{(11x + 4)} - 2x$

11) $y = 3\sqrt{-2(4x + 8)} + 5$

12) $y = \sqrt[5]{(3x^2 - 12)} - 6$

13) $y = 3\sqrt{x - 5} - 2$

14) $y = \sqrt[3]{6x + 9} - 4$

✍ **Sketch the graph of each function.**

15) $y = -3\sqrt{x} + 5$

16) $y = 3\sqrt{x} - 6$

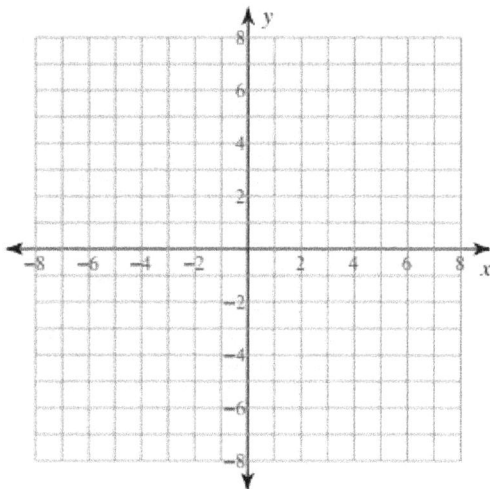

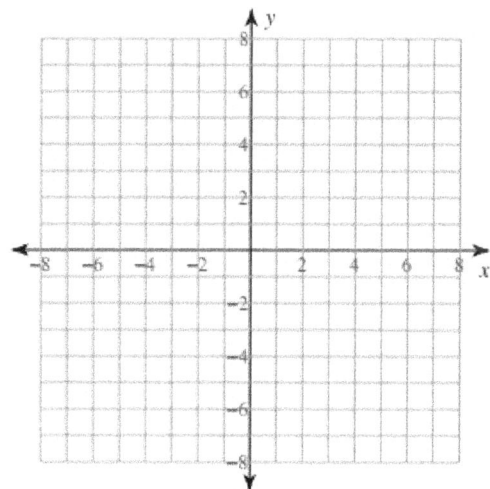

Solving Radical Equations

✍ **Solve each equation. Remember to check for extraneous solutions.**

1) $\sqrt{a} = 9$

2) $\sqrt{v} = 6$

3) $\sqrt{r} = 4$

4) $8 = 16\sqrt{x}$

5) $\sqrt{x+3} = 18$

6) $6 = \sqrt{x-7}$

7) $4 = \sqrt{r-3}$

8) $\sqrt{x-5} = 7$

9) $12 = \sqrt{x-4}$

10) $\sqrt{m+5} = 8$

11) $7\sqrt{5a} = 35$

12) $6\sqrt{2x} = 48$

13) $2 = \sqrt{6x-32}$

14) $\sqrt{304-4x} = 4$

15) $\sqrt{r+2} - 8 = 6$

16) $-21 = -7\sqrt{r+9}$

17) $60 = 6\sqrt{5v}$

18) $x = \sqrt{40-3x}$

19) $\sqrt{90-27a} = 3a$

20) $\sqrt{-8n+88} = 4$

21) $\sqrt{15r-5} = 4r-3$

22) $\sqrt{-64+32x} = 4x$

23) $\sqrt{4x+15} = \sqrt{2x+11}$

24) $\sqrt{12v} = \sqrt{15v-21}$

25) $\sqrt{9-x} = \sqrt{x-3}$

26) $\sqrt{6m+34} = \sqrt{8m+34}$

27) $\sqrt{7r+32} = \sqrt{-8-3r}$

28) $\sqrt{4k+10} = \sqrt{2-4k}$

29) $-20\sqrt{x-13} = -40$

30) $\sqrt{90-2x} = \sqrt{\dfrac{x}{4}}$

Answers of Worksheets

Evaluating Function

1) $h(x) = -8x + 3$
2) $k(a) = 2a - 14$
3) $d(t) = 11t$
4) $f(x) = \frac{5}{12}x - \frac{7}{12}$
5) $m(n) = 24n - 210$
6) $c(p) = p^2 - 5p + 10$
7) -13
8) 14
9) -3
10) 22
11) 9
12) 26
13) 2
14) 7
15) -27
16) 8.2
17) 15
18) 15
19) 74
20) 75
21) $-1\frac{7}{8}$
22) $-3\frac{8}{9}$
23) 87
24) -28
25) 14
26) $-\frac{14b+9}{3b}$
27) $12a - 50$
28) $-2x + 18$
29) $2x^2 + 16$
30) $16x^4 - 20$

Adding and Subtracting Functions

1) -2
2) 1
3) -4
4) $-3x^2 - 6x - 19$
5) -43
6) 15
7) $-7\frac{2}{3}$
8) $4a^2 + 8a - 4$
9) $-18x^2 - 18x - 12$
10) $-2t^2 - 11t + 1$
11) $5x^4 - 5x^2 + 9$
12) $-81x^8 + 22$

Multiplying and Dividing Functions

1) -55
2) -70
3) 153
4) -1
5) $4\frac{4}{7}$
6) 2
7) 84
8) 2
9) 0
10) -62
11) $6x^4 - 17x^3 + 5x^2 + 3x - 1$
12) $5x^8 - 2x^6 - 10x^4 + 4x^2$

Composition of Functions

1) -13
2) -1
3) 26
4) -10
5) -17
6) 10
7) $\frac{25}{8}$
8) $\frac{1}{8}$
9) 8
10) $\frac{5}{8}$
11) $\frac{49}{8}$
12) $-\frac{1}{2}(x^2 - \frac{3}{2})$
13) 1
14) -2
15) 2
16) 6
17) -1
18) $-2x$

19) 3   21) 2   23) −7

20) −9   22) $\sqrt{13}$ 24) −18

Quadratic Equations

1) $x^2 + 2x - 24$
2) $x^2 + 12x + 35$
3) $x^2 + 2x - 48$
4) $x^2 - 7x - 18$
5) $x^2 - 15x + 56$
6) $3x^2 - 7x - 6$
7) $4x^2 + 5x - 6$
8) $4x^2 - x - 5$
9) $7x^2 - 41x - 6$
10) $15x^2 - 12x - 3$

11) $(x-4)(x+2)$
12) $(x+5)(x+3)$
13) $(x-6)(x+4)$
14) $(x-3)(x-7)$
15) $(x+3)(x+7)$
16) $(4x+5)(x+1)$
17) $(5x-2)(x+3)$
18) $(5x-3)(x+4)$
19) $(2x+5)(x+1)$
20) $3(x-2)(3x-1)$

21) $x = -6, x = 3$
22) $x = -1, x = -8$
23) $x = -2, x = -5$
24) $x = 1, x = -2$
25) $x = 3, x = -4$
26) $x = -3, x = -8$
27) $x = -4, x = -\frac{1}{2}$
28) $x = 4, x = -7$
29) $x = 3, x = -4$
30) $x = -3, x = 2$

Solving quadratic equations

1) $\{-9, 1\}$
2) $\{-6, -7\}$
3) $\{8, -3\}$
4) $\{6, 4\}$
5) $\{-2, -12\}$
6) $\{-\frac{4}{5}, -7\}$
7) $\{-\frac{5}{4}, -\frac{1}{6}\}$
8) $\{-\frac{7}{2}, -8\}$

9) $\{-6, -5\}$
10) $\{-\frac{1}{6}, -8\}$
11) $\{8, 0\}$
12) $\{4, -4\}$
13) $\{2, 1\}$
14) $\{-4, -1\}$
15) $\{1, -9\}$
16) $\{2, -12\}$

17) $\{2, -8\}$
18) $\{-3, -6\}$
19) $\{-4, -9\}$
20) $\{5, -3\}$
21) $\{-5, -3\}$
22) $\{1, 3\}$
23) $\{\frac{6}{5}, \frac{3}{2}\}$
24) $\{\frac{6}{7}, 0\}$

25) $\{-\frac{7}{2}, 2\}$
26) $\{\frac{3}{5}, 2\}$
27) $\{-\frac{4}{3}, -4\}$
28) $\{-8, -7\}$
29) $\{4, -1\}$
30) $\{3, -6\}$
31) $\{3, 8\}$
32) $\{\frac{15}{17}, 0\}$

Quadratic formula and the discriminant

1) 576
2) 68
3) 8
4) −11
5) 28

6) 116
7) −11
8) 52
9) 40
10) 69

11) 64
12) 169
13) 336
14) 33
15) 316

16) 132
17) 104
18) 16
19) 1
20) 33

21) 0, *one real solution*
22) 0, *one real solution*
23) −783, *no solution*

24) $0, one\ real\ solution$

26) $0, one\ real\ solution$

28) $0, one\ real\ solution$

25) $-127, no\ solution$

27) $-703, no\ solution$

Graphing quadratic functions

1) $(-3, 2), x = -3$

2) $(3, -2), x = 3$

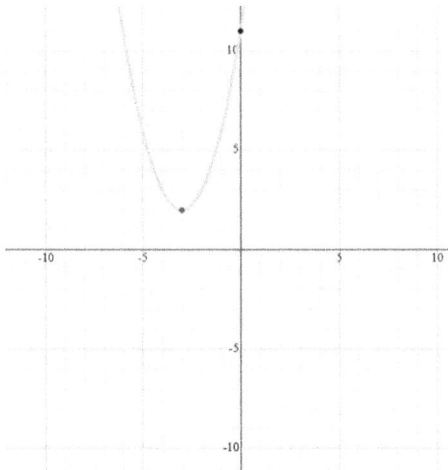

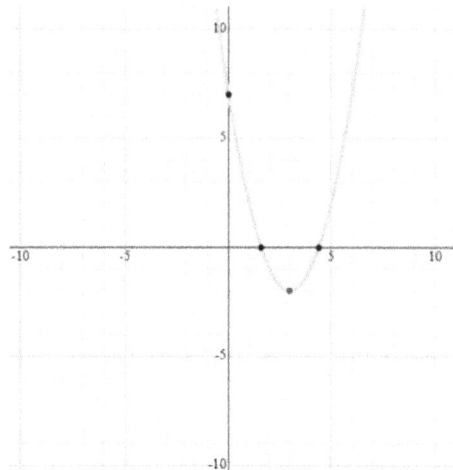

3) $(4, 6), x = 4$

4) $(-1, 12), x = -1$

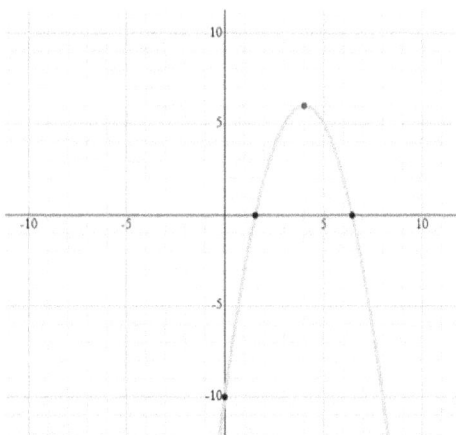

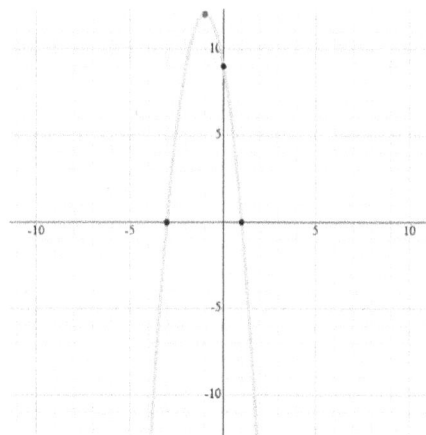

Quadratic inequalities

1) $-5 < x < 5$

7) no solution

13) no solution

2) $-4 < x < -2$

8) $-2 \leq x \leq 7$

14) $x = 6$

3) no solution

9) $x \leq -1\ or\ x \geq 7$

15) $x \leq 7\ or\ x \geq 9$

4) $x < -4\ or\ x > -4$

10) $-9 < x < 1$

16) $x \leq 5\ or\ x \geq 11$

5) $x \leq -1\ or\ x \geq 10$

11) $x < -12\ or\ x > 6$

17) $-9 \leq x \leq 9$

6) $x < -5\ or\ x > -5$

12) $-3 < x < 4$

18) $x \leq 3\ or\ x \geq 14$

19) no solution

20) $x < -3$ or $x > 4$

21) no solution

22) $x \leq 1$ or $x \geq 5$

23) $x < -3$ or $x > 5$

24) $x \geq 3$

25) $x < -6$ or $x > 3$

26) $-6 < x < 1$

27) no solution

28) $x \leq -1$ or $x \geq 6$

29) $-5 < x < -3$

30) no solution

31) $x \leq -4$ or $x \geq 3$

32) $x \leq -1$ or $x \geq 3$

Domain and range of radical functions

1) domain: $x \geq -8$

   range: $y \geq -7$

2) domain: {all real numbers}

   range: {all real numbers}

3) domain: $x \geq 3$

   range: $y \geq 3$

4) domain: {all real numbers}

   range: {all real numbers}

5) domain: $x \geq -5$

   range: $y \geq 6$

6) domain: {all real numbers}

   range: {all real numbers}

7) domain: {all real numbers}

   range: $y \geq 8\sqrt{2} + 3$

8) domain: {all real numbers}

   range: {all real numbers}

9) domain: $x \geq -2$

   range: $y \geq -3$

10) domain: {all real numbers}

   range: {all real numbers}

11) domain: $x \leq -2$

   range: $y \geq 5$

12) domain: {all real numbers}

   range: {all real numbers}

13) domain: $x \geq 5$

   range: $y \geq -2$

14) domain: {all real numbers}

   range: {all real numbers}

15)

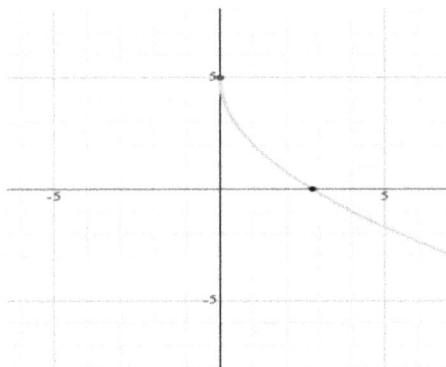

16)

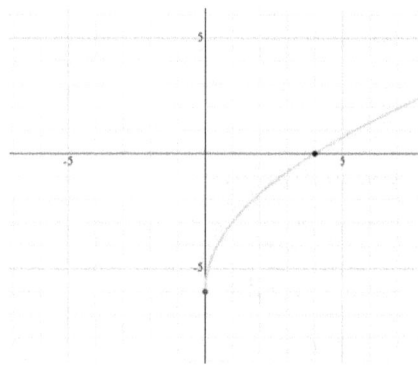

Solving radical equations

1) {81}

2) {36}

3) {16}

4) $\{\frac{1}{4}\}$

5) {321}

6) {43}

7) {19}

8) {54}

9) {148}

10) {59}

11) {5}

12) {32}

13) {6}

14) {72}

15) {194}

16) {0}

17) {20}

18) {5}

19) {2}

20) {9}

21) {2}

22) no solution

23) {−2}

24) {7}

25) {6}

26) {0}

27) {−4}

28) {−1}

29) {17}

30) {40}

Chapter 11 :

Sequences and Series

Topics that you'll practice in this chapter:

✓ Arithmetic Sequences

✓ Geometric Sequences

✓ Comparing Arithmetic and Geometric Sequences

✓ Finite Geometric Series

✓ Infinite Geometric Series

"The fact that people will be full of greed, fear, or folly is predictable. The sequence is not predictable." -Warren Buffett

Arithmetic Sequences

✍ **Find the next three terms of each arithmetic sequence.**

1) $32, 26, 20, 14, 8, \ldots$

2) $-56, -44, -32, -20, \ldots$

3) $17, 26, 35, 44, 53, \ldots$

4) $5, 11, 17, 23, 29, \ldots$

✍ **Given the first term and the common difference of an arithmetic sequence find the first five terms and the explicit formula.**

5) $a_1 = 20, d = 3$ 7) $a_1 = 32, d = 6$

6) $a_1 = -11, d = -5$ 8) $a_1 = 240, d = -80$

✍ **Given a term in an arithmetic sequence and the common difference find the first five terms and the explicit formula.**

9) $a_{20} = -500, d = -50$ 11) $a_{51} = -88.2, d = -5.2$

10) $a_{24} = 98, d = 7$ 12) $a_{68} = -980, d = -27$

✍ **Given a term in an arithmetic sequence and the common difference find the recursive formula and the three terms in the sequence after the last one given.**

13) $a_{21} = -187, d = -9$ 15) $a_{31} = 58.2, d = 1.8$

14) $a_{12} = 63.5, d = 5.2$ 16) $a_{42} = 6.8, d = 0.4$

Geometric Sequences

✍ **Determine if the sequence is geometric. If it is, find the common ratio.**

1) $1, -7, 49, -343, \ldots$

3) $8, 24, 48, 240, \ldots$

2) $-3, -12, -48, -192, \ldots$

4) $-5, -10, -20, -40, \ldots$

✍ **Given the first term and the common ratio of a geometric sequence find the first five terms and the explicit formula.**

5) $a_1 = 0.4, r = -3$

6) $a_1 = 0.2, r = 4$

✍ **Given the recursive formula for a geometric sequence find the common ratio, the first five terms, and the explicit formula.**

7) $a_n = a_{n-1} \times 4, a_1 = 2$

9) $a_n = a_{n-1} \cdot 5, a_1 = 0.2$

8) $a_n = a_{n-1} \cdot (-2), a_1 = -4$

10) $a_n = a_{n-1} \cdot 3, a_1 = -3$

✍ **Given two terms in a geometric sequence find the 6th term and the recursive formula.**

11) $a_3 = 576$ and $a_5 = 36$

12) $a_2 = -0.4$ and $a_4 = -1.6$

Comparing Arithmetic and Geometric Sequences

✎ **For each sequence, state if it is arithmetic, geometric, or neither.**

1) $6, 11, 16, 21, \ldots$

2) $2, 5, 8, 11, \ldots$

3) $3, 10, 20, 27, \ldots$

4) $1, 11, 22, 33, 44, \ldots$

5) $4, 8, 10, 24, 96, \ldots$

6) $2, 10, 20, 50, 200, \ldots$

7) $0.2, 1, 5, 25, 125, \ldots$

8) $4, 12, 36, 108, \ldots$

9) $-27, -34, -41, -48, -55, \ldots$

10) $-3, 15, -75, 375, -1875, \ldots$

11) $10, 25, 50, 65, 80, \ldots$

12) $5, 15, 150, 250, 350 \ldots$

13) $-35, -20, -5, 10, 25, \ldots$

14) $a_n = 4 \cdot 8^{n-1}$

15) $a_n = 3 \cdot 6^{n-1}$

16) $a_n = 8 - 4n$

17) $a_n = -210 + 310n$

18) $a_n = 53 + 53n$

19) $a_n = -10 \cdot (-5)^{n-1}$

20) $a_n = 49 + 63n$

21) $a_n = (3n)^4$

22) $a_n = 40 + 8n$

23) $a_n = -(13)^{n-1}$

24) $a_n = -10 \cdot (0.2)^{n-1}$

25) $a_n = \frac{3n+2}{4^n}$

26) $a_n = \frac{47+13n}{7^n}$

27) $a_n = \frac{12-12n}{12^n}$

28) $a_n = \frac{32-a_{n-1}}{2n}$

29) $a_n = \frac{4}{15} - \frac{2}{7}n$

Finite Geometric Series

✍ **Evaluate the related series of each sequence.**

1) $-2, 4, -8, 16$

2) $-1, 6, -36, 216, -1,296$

3) $-2, 8, -32, 128, -512$

4) $4, 8, 16, 32, 64$

5) $-4, -12, -36, -108$

6) $5, -15, 45, -135, 405$

✍ **Evaluate each geometric series described.**

7) $1 + 5 + 25 + 125 \ldots, n = 5$ _____

8) $1 - 6 + 36 - 216 \ldots, n = 6$ _____

9) $-2 - 6 - 18 - 54 \ldots, n = 8$ _____

10) $0.2 - 1 + 5 - 25 \ldots, n = 6$ _____

11) $0.6 - 3.6 + 21.6 - 129.6 \ldots, n = 5$ _____

12) $-1 - 4 - 16 - 64 \ldots, n = 6$ _____

13) $a_1 = -2, r = 6, n = 5$ _____

14) $a_1 = 1, r = 9, n = 6$ _____

15) $\sum_{n=1}^{6} 4 \cdot (-3)^{n-1}$ _____

16) $\sum_{n=1}^{7} 2 \cdot (-5)^{n-1}$ _____

17) $\sum_{n=1}^{10} 0.1 \cdot (2)^{n-1}$ _____

18) $\sum_{m=1}^{6} (-4)^{m-1}$ _____

19) $\sum_{m=1}^{6} 5 \times (3)^{m-1}$ _____

20) $\sum_{k=1}^{5} 7 \times (6)^{k-1}$ _____

Infinite Geometric Series

✎ **Determine if each geometric series converges or diverges.**

1) $a_1 = -1.8, \ r = 6$

2) $a_1 = 10.8, r = 0.3$

3) $a_1 = -2, r = 6.1$

4) $a_1 = 5, r = 0.24$

5) $a_1 = 1.2, r = 8$

6) $-1, 6, -36, 216, \ldots$

7) $6, -1, \dfrac{1}{6}, -\dfrac{1}{36}, \dfrac{1}{216}, \ \ldots$

8) $512 + 64 + 8 + 1 \ \ldots$

9) $-5 + \dfrac{15}{7} - \dfrac{45}{49} + \dfrac{135}{343} \ \cdots$

10) $\dfrac{400}{459} - \dfrac{200}{153} + \dfrac{100}{51} - \dfrac{50}{17} \ \cdots$

✎ **Evaluate each infinite geometric series described.**

11) $a_1 = 2, r = -\dfrac{1}{4}$

12) $a_1 = 36, r = -\dfrac{1}{6}$

13) $a_1 = 9, r = \dfrac{1}{3}$

14) $a_1 = 12, r = \dfrac{1}{7}$

15) $1 + 0.2 + 0.04 + 0.008 + \cdots$

16) $64 - 16 + 4 - 1 \ \ldots,$

17) $1 - 0.3 + 0.09 - 0.027 \ \ldots,$

18) $-5 + \dfrac{15}{7} - \dfrac{45}{49} + \dfrac{135}{343} \ \cdots,$

19) $\sum_{k=1}^{\infty} 7^{k-1}$

20) $\sum_{i=1}^{\infty} \left(\dfrac{2}{5}\right)^{i-1}$

21) $\sum_{k=1}^{\infty} \left(-\dfrac{2}{9}\right)^{k-1}$

22) $\sum_{n=1}^{\infty} 6\left(\dfrac{1}{3}\right)^{n-1}$

Answers of Worksheets

Arithmetic Sequences

1) $2, -4, -10$

2) $-8, 4, 16$

3) $62, 71, 80$

4) $35, 41, 47$

5) First Five Terms: $20, 23, 26, 29, 32$, Explicit: $a_n = 20 + 3(n - 1)$

6) First Five Terms: $-11, -16, -21, -26, -31$, Explicit: $a_n = -11 - 5(n - 1)$

7) First Five Terms: $32, 38, 44, 50, 56$, Explicit: $a_n = 32 + 6(n - 1)$

8) First Five Terms: $240, 160, 80, 0, -80$, Explicit: $a_n = 240 - 80(n - 1)$

9) First Five Terms: $450, 400, 350, 300, 250$, Explicit: $a_n = 450 - 50(n - 1)$

10) First Five Terms: $-63, -56, -49, -42, -35$, Explicit: $a_n = -63 + 7(n - 1)$

11) First Five Terms: $171.8, 166.6, 161.4, 156.2, 151$, Explicit: $a_n = 171.8 - 5.2(n - 1)$

12) First Five Terms: $829, 802, 775, 748, 721$, Explicit: $a_n = 829 - 27(n - 1)$

13) Next 3 terms: $-196, -205, -214$, Recursive: $a_n = a_{n-1} - 9, a_1 = -7$

14) Next 3 terms: $68.7, 73.9, 79.1, 84.3$ Recursive: $a_n = a_{n-1} + 5.2, \ a_1 = 6.3$

15) Next 3 terms: $60, 61.8, 63.6$, Recursive: $a_n = a_{n-1} + 1.8, a_1 = 4.2$

16) Next 3 terms: $7.2, 7.6, 8$, Recursive: $a_n = a_{n-1} + 0.4, a_1 = -9.6$

Geometric Sequences

1) $r = -7$

2) $r = 4$

3) not geometric

4) $r = 2$

5) First Five Terms: $0.4, -1.2, 3.6, -10.8, 32.4$

   Explicit: $a_n = 0.4 \times (-3)^{n-1}$

6) First Five Terms: $0.2, 0.8, 3.2, 12.8, 51.2$

   Explicit: $a_n = 0.2 \times (4)^{n-1}$

7) Common Ratio: $r = 4$

   First Five Terms: $2, 8, 32, 128, 512$

   Explicit: $a_n = 2 . (4)^{n-1}$

8) Common Ratio: $r = -2$

First Five Terms: $-4, 8, -16, 32, -64$

Explicit: $a_n = -4 \cdot (-2)^{n-1}$

9) Common Ratio: $r = 5$

First Five Terms: $0.2, 1, 5, 25, 125, 625$

Explicit: $a_n = 0.2 \cdot (5)^{n-1}$

10) Common Ratio: $r = 3$

First Five Terms: $-3, -9, -27, -81, -243$

Explicit: $a_n = -3 \cdot (3)^{n-1}$

11) $a_6 = -9$, Recursive: $a_n = a_{n-1} \cdot (\frac{-1}{4})$, $a_1 = 9,216$

12) $a_6 = -6.4$, Recursive: $a_n = a_{n-1} \cdot (-2)$, $a_1 = 0.2$

Comparing Arithmetic and Geometric Sequences

1) Arithmetic
2) Arithmetic
3) Neither
4) Neither
5) Neither
6) Neither
7) Geometric
8) Geometric
9) Arithmetic
10) Geometric

11) Neither
12) Neither
13) Arithmetic
14) Geometric
15) Geometric
16) Arithmetic
17) Arithmetic
18) Arithmetic
19) Geometric
20) Arithmetic

21) Neither
22) Arithmetic
23) Geometric
24) Geometric
25) Neither
26) Neither
27) Neither
28) Neither
29) Arithmetic

Finite Geometric

1) 10
2) $-1,111$
3) -410
4) 124
5) -160
6) 305
7) 781

8) $-6,665$
9) $-6,560$
10) -520.8
11) 666.6
12) $-1,365$
13) $-3,110$
14) 66,430

15) -728
16) 26,042
17) 102.3
18) -819
19) 1,820
20) 10,885

Infinite Geometric

1) Diverges

2) Converges

3) Diverges

4) Converges

5) Diverges

6) Diverges

7) Converges

8) Converges

9) Converges

10) Converges

11) $\frac{8}{5}$

12) $\frac{216}{7}$

13) $\frac{27}{2}$

14) 14

15) $\frac{5}{4}$

16) $\frac{256}{5}$

17) $\frac{3}{4}$

18) $-\frac{7}{2}$

19) Infinite

20) $\frac{5}{3}$

21) $\frac{9}{11}$

22) 9

Chapter 12 :
Logarithms

Topics that you'll practice in this chapter:

- ✓ Rewriting Logarithms
- ✓ Evaluating Logarithms
- ✓ Properties of Logarithms
- ✓ Natural Logarithms
- ✓ Exponential Equations Requiring Logarithms
- ✓ Solving Logarithmic Equations

Mathematics is an art of human understanding.
— William Thurston

Rewriting Logarithms

✍ **Rewrite each equation in exponential form.**

1) $\log_3 27 = 3$

2) $\log_2 128 = 7$

3) $\log_6 1{,}296 = 4$

4) $\log_5 625 = 4$

5) $\log_{11} 121 = 2$

6) $\log_{12} 1{,}728 = 3$

7) $\log_9 729 = 3$

8) $\log_3 729 = 6$

9) $\log_{10} 10{,}000 = 4$

10) $\log_7 343 = 3$

11) $\log_4 1{,}024 = 5$

12) $\log_{12} 144 = 2$

13) $\log_{13} 2{,}197 = 3$

14) $\log_{25} 5 = \frac{1}{2}$

15) $\log_{81} 3 = \frac{1}{4}$

16) $\log_{3{,}125} 5 = \frac{1}{5}$

17) $\log_{1{,}000} 10 = \frac{1}{3}$

18) $\log_5 \frac{1}{125} = -3$

19) $\log_4 \frac{1}{16} = -2$

20) $\log_a \frac{7}{4} = b$

✍ **Rewrite each exponential equation in logarithmic form.**

21) $2^5 = 32$

22) $4^3 = 64$

23) $5^4 = 625$

24) $11^3 = 1{,}331$

25) $3^5 = 243$

26) $6^4 = 1{,}296$

27) $7^4 = 2{,}401$

28) $9^3 = 729$

29) $4^{-5} = \frac{1}{1{,}024}$

30) $3^{-8} = \frac{1}{6{,}561}$

31) $11^{-2} = \frac{1}{121}$

32) $12^{-3} = \frac{1}{1{,}728}$

33) $4^{-5} = \frac{1}{1{,}024}$

34) $10^{-5} = \frac{1}{100{,}000}$

Evaluating Logarithms

✍ **Evaluate each logarithm.**

1) $\log_3 729 =$

2) $\log_2 256 =$

3) $\log_3 243 =$

4) $\log_4 64 =$

5) $\log_8 64 =$

6) $\log_{11} 121 =$

7) $\log_{10} 10,000 =$

8) $\log_5 \frac{1}{25} =$

9) $\log_4 \frac{1}{256} =$

10) $\log_2 \frac{1}{32} =$

11) $\log_6 \frac{1}{36} =$

12) $\log_9 \frac{1}{81} =$

13) $\log_{12} \frac{1}{144} =$

14) $\log_{1,000} \frac{1}{10} =$

15) $\log_{243} 3 =$

16) $\log_4 \frac{1}{16} =$

17) $\log_8 \frac{1}{512} =$

18) $\log_3 \frac{1}{81} =$

✍ **Circle the points which are on the graph of the given logarithmic functions.**

19) $y = 4\log_4(3x - 2) + 1$ $(3,4),$ $(2,5),$ $(7,4)$

20) $y = 5\log_6(12x) - 7$ $(2,-2),$ $(\frac{1}{3},12),$ $(\frac{1}{2},-2)$

21) $y = -2\log_3 9(x - 5) + 5$ $(6,-3),$ $(8,-1),$ $(1,6)$

22) $y = \frac{1}{4}\log_6(6x) + \frac{1}{2}$ $(6,1),$ $(6,\frac{1}{4}),$ $(4,\frac{1}{4})$

23) $y = -2\log_8 8(x + 4) + 9$ $(-4,0),$ $(0,9),$ $(-2,6\frac{1}{3})$

24) $y = -\log_5(x + 15) - 6$ $(10,-\frac{1}{5}),$ $(10,-8),$ $(11,-\frac{2}{5})$

25) $y = -3\log_2(2x + 6) + 7$ $(5,-5),$ $(-5,-5),$ $(-2,-2)$

Properties of Logarithms

✎ **Expand each logarithm.**

1) $\log(11 \times 4) =$

2) $\log(13 \times 5) =$

3) $\log(4 \times 12) =$

4) $\log\left(\frac{2}{7}\right) =$

5) $\log\left(\frac{4}{9}\right) =$

6) $\log\left(\frac{5}{8}\right)^3 =$

7) $\log(6 \times 5^4) =$

8) $\log\left(\frac{14}{3}\right)^5 =$

9) $\log\left(\frac{3^4}{8}\right) =$

10) $\log(x \times y)^8 =$

11) $\log(x^6 \times y^{12} \times z^2) =$

12) $\log\left(\frac{u^8}{v^3}\right) =$

13) $\log\left(\frac{x}{y^7}\right) =$

✎ **Condense each expression to a single logarithm.**

14) $\log 8 - \log 13 =$

15) $\log 6 + \log 11 =$

16) $4\log 2 - 7\log 5 =$

17) $10\log 4 - 3\log 7 =$

18) $3\log 9 - \log 17 =$

19) $11\log 6 - 9\log 4 =$

20) $\log 15 - 6\log 7 =$

21) $6\log 8 + 4\log 10 =$

22) $12\log 5 + 14\log 9 =$

23) $17\log_8 a + 6\log_8 b =$

24) $2\log_9 x - 3\log_9 y =$

25) $\log_{11} u - 16\log_{11} v =$

26) $8\log_{15} u + 9\log_{15} v =$

27) $32\log_6 u - 25\log_6 v =$

Natural Logarithms

✍ **Solve each equation for** x.

1) $e^x = 9$

2) $e^x = 36$

3) $e^x = 49$

4) $\ln x = 3$

5) $\ln(\ln x) = 7$

6) $e^x = 4$

7) $\ln(5x + 2) = 1$

8) $\ln(7x + 4) = 3$

9) $\ln(9x + 5) = 4$

10) $\ln x = \frac{1}{9}$

11) $\ln 11x = e^5$

12) $\ln x = \ln 6 + \ln 7$

13) $\ln x = 4\ln 3 + \ln 2$

✍ **Evaluate without using a calculator.**

14) $11\ln e =$

15) $\ln e^{10} =$

16) $4 \ln e =$

17) $\ln e^{21} =$

18) $32\ln e =$

19) $4\ln e^5 =$

20) $e^{\ln 22} =$

21) $e^{3\ln 3} =$

22) $e^{3\ln 5} =$

23) $\ln \sqrt[11]{e} =$

✍ **Reduce the following expressions to simplest form.**

24) $e^{-4\ln 9 + 4\ln 3} =$

25) $e^{-3\ln\left(\frac{5}{4e}\right)} =$

26) $2\ln(e^4) =$

27) $\ln\left(\frac{1}{e}\right)^4 =$

28) $e^{\ln 9 + 3\ln 3} =$

29) $e^{\ln\left(\frac{13}{e}\right)} =$

30) $8\ln(1^{-3e}) =$

31) $2\ln\left(\frac{1}{e}\right)^{-3} =$

32) $6\ln\left(\frac{\sqrt[3]{e}}{3e}\right) =$

33) $e^{-4\ln e + 2\ln 5} =$

34) $e^{\ln\frac{4}{e}} =$

35) $11\ln(e^e) =$

Exponential Equations and Logarithms

✍ **Solve each equation for the unknown variable.**

1) $3^{4n} = 243$

2) $5^{3r} = 625$

3) $6^{2n-1} = 216$

4) $16^{2r+3} = 4$

5) $169^{2x} = 13$

6) $7^{-3v-3} = 49$

7) $2^{4n} = 128$

8) $11^{n-1} = 1,331$

9) $\frac{9^{3a}}{3^{2a}} = 729$

10) $13^5 \times 13^{-4v} = 169$

11) $4^{3n} = \frac{1}{64}$

12) $(\frac{1}{11})^{2n} = 121$

13) $2,187^{3x} = 3$

14) $13^{5-7x} = 13^{-2x}$

15) $11^{-3x} = 11^{2x-7}$

16) $3^{5n} = 243$

17) $17^{5x+3} = 17^{6x}$

18) $15^{3n} = 225$

19) $4^{-3k} = 512$

20) $8^{-4r} = 8^{-5r+2}$

21) $8^{2x+3} = 8^{5x}$

22) $10^{3x-2} = 100,000$

23) $16 \times 64^{-v} = 128$

24) $\frac{128}{2^{-3m}} = 2^{4m+5}$

25) $14^{-5n} \times 14^{2n+3} = 14^{-2n}$

26) $(\frac{1}{9})^{4n+3} \times (\frac{1}{9})^{-3n-8} = (\frac{1}{9})^{-4n}$

✍ **Solve each problem. (Round to the nearest whole number)**

27) A substance decays 16% each day. After 8 days, there are 6 milligrams of the substance remaining. How many milligrams were there initially? _____

28) A culture of bacteria grows continuously. The culture doubles every 4 hours. If the initial number of bacteria is 20, how many bacteria will there be in 13 hours?

_____

29) Bob plans to invest $11,200 at an annual rate of 3.5%. How much will Bob have in the account after three years if the balance is compounded quarterly? _____

30) Suppose you plan to invest $8,000 at an annual rate of 5%. How much will you have in the account after 6 years if the balance is compounded monthly? _____

Solving Logarithmic Equations

✍ **Find the value of the variables in each equation.**

1) $2\log(x) + 5 = 9$

2) $\log_4 4x + 3 = 5$

3) $-\log_8(8x) + 2 = 3$

4) $\log 2x - \log 4 = 1$

5) $\log 5x + \log 25 = 1$

6) $\log 4 - \log x = 3$

7) $\log 4x + \log 2 = \log 16$

8) $-6\log_3(5x - 1) = -36$

9) $\log 4x = \log(8x - 1)$

10) $\log(4k - 6) = \log(k - 3)$

11) $\log(5p + 2) = \log(p + 4)$

12) $-30 + \log_4(3n + 2) = -30$

13) $\log_4(4x - 4) = \log_4(x^2)$

14) $\log_8(k^2 + 15) = \log_8(-6k - 3)$

15) $\log(16 + 6b) = \log(10b^2 + 12b)$

16) $\log_6(2x + 5) - \log_6 x = \log_6 9$

17) $\log_5 5 + \log_5(x^2 + 1) = \log_5 25$

18) $\log_6(x + 3) + \log_6(x + 1) = \log_6 8$

✍ **Find the value of x in each natural logarithm equation.**

19) $\ln 8 - \ln(4x + 8) = 4$

20) $\ln(x + 5) - \ln(x + 2) = \ln 10$

21) $\ln e^6 - \ln(x - 1) = 3$

22) $\ln(2x - 8) + \ln(x - 4) = \ln 8$

23) $\ln 5x - \ln(x + 4) = \ln 2$

24) $\ln(8x - 4) - \ln(x - 2) = \ln 25$

25) $\ln(3x + 2) - 4\ln 2 = 5$

26) $\ln(2x - 5) + \ln(x - 3) = \ln 6$

27) $\ln(x - 1) + \ln(4x - 7) = \ln(7)$

28) $2\ln 3x - \ln(x + 10) = \ln 2x$

29) $\ln x^4 + \ln x^8 = 4\ln(2x)$

30) $\ln x^{10} - \ln(x^2 + 10) = 10\ln 2x$

31) $8\ln(x - 2) = 4\ln(x^2 - 4x + 4)$

32) $\ln(x^4 + 10) = \ln(x^2 + 9)$

33) $2\ln x - 2\ln(x + 8) = \ln(x^2)$

34) $\ln(2x + 1) - \ln(4x + 1) = \ln 4$

35) $\ln 16 + 2\ln(x - 2) = \ln 4$

36) $\ln e^2 + \ln(5x - 6) = \ln(5) + 3$

Answers of Worksheets

Rewriting Logarithms

1) $3^3 = 27$

2) $2^7 = 128$

3) $6^4 = 1,296$

4) $5^4 = 625$

5) $11^2 = 121$

6) $12^3 = 1,728$

7) $9^3 = 729$

8) $3^6 = 729$

9) $10^4 = 10,000$

10) $7^3 = 343$

11) $4^5 = 1,024$

12) $12^2 = 144$

13) $13^3 = 2,197$

14) $25^{\frac{1}{2}} = 5$

15) $81^{\frac{1}{4}} = 3$

16) $3,125^{\frac{1}{5}} = 5$

17) $1,000^{\frac{1}{3}} = 10$

18) $5^{-3} = \frac{1}{125}$

19) $4^{-2} = \frac{1}{16}$

20) $a^b = \frac{7}{4}$

21) $\log_2 32 = 5$

22) $\log_4 64 = 3$

23) $\log_5 625 = 4$

24) $\log_{11} 1,331 = 3$

25) $\log_3 243 = 5$

26) $\log_6 1,296 = 4$

27) $\log_7 2,401 = 4$

28) $\log_9 729 = 3$

29) $\log_4 \frac{1}{1,024} = -5$

30) $\log_3 \frac{1}{6,561} = -8$

31) $\log_{11} \frac{1}{121} = -2$

32) $\log_{12} \frac{1}{1,728} = -3$

33) $\log_4 \frac{1}{1,024} = -5$

34) $\log_{10} \frac{1}{100,000} = -5$

Evaluating Logarithms

1) 6

2) 8

3) 5

4) 3

5) 2

6) 2

7) 4

8) -2

9) -4

10) -5

11) -2

12) -2

13) -2

14) $-\frac{1}{3}$

15) $\frac{1}{5}$

16) -2

17) -3

18) -4

19) $(2, 5)$

20) $\left(\frac{1}{2}, -2\right)$

21) $(8, -1)$

22) $(6, 1)$

23) $\left(-2, 6\frac{1}{3}\right)$

24) $(10, -8)$

25) $(5, -5)$

Properties of Logarithms

1) $\log 11 + \log 4$

2) $\log 13 + \log 5$

3) $\log 4 + \log 12$

4) $\log 2 - \log 7$

5) $\log 4 - \log 9$

6) $3 \log 5 - 3 \log 8$

7) $\log 6 + 4 \log 5$

8) $5\log 14 - 5 \log 3$

9) $4 \log 3 - \log 8$

10) $8 \log x + 8 \log y$

11) $6 \log x + 12 \log y + 2 \log z$

12) $8 \log u - 3 \log v$

13) $\log x - 7 \log y$

14) $\log \frac{8}{13}$

15) $\log(6 \times 11)$

16) $\log \frac{2^4}{5^7}$

17) $\log \frac{4^{10}}{7^3}$

18) $\log \frac{9^3}{17}$

19) $\log \frac{6^{11}}{4^9}$

20) $\log \frac{15}{7^6}$

21) $\log (8^6 \times 10^4)$

22) $\log (5^{12} \times 9^{14})$

23) $\log_8 (a^{17} b^6)$

24) $\log_9 \frac{x^2}{y^3}$

25) $\log_{11} \frac{u}{v^{16}}$

26) $\log_{15} (u^8 \times v^9)$

27) $\log_6 \frac{u^{32}}{v^{25}}$

Natural Logarithms

1) $x = \ln 9$

2) $x = \ln 36, x = 2\ln (6)$

3) $x = \ln 49, x = 2\ln (7)$

4) $x = e^3$

5) $x = e^{e^7}$

6) $x = \ln 4$

7) $x = \frac{e-2}{5}$

8) $x = \frac{e^3 - 4}{7}$

9) $x = \frac{e^4 - 5}{9}$

10) $x = \sqrt[9]{e}$

11) $x = \frac{e\, e^5}{11}$

12) $x = 42$

13) $x = 162$

14) 11

15) 10

16) 4

17) 21

18) 32

19) 20

20) 22

21) 27

22) 125

23) $\frac{1}{11}$

24) $\frac{1}{81}$

25) $\frac{64e^3}{125}$

26) 8

27) -4

28) 243

29) $\frac{13}{e}$

30) 0

31) 6

32) $\ln \left(\frac{1}{3^6 e^4}\right) = -10.6$

33) $25e^{-4} = \frac{25}{e^4}$

34) $\frac{4}{e}$

35) $11e$

Exponential Equations and Logarithms

1) $\frac{5}{4}$

2) $\frac{4}{3}$

3) 2

4) $-\frac{5}{4}$

5) $\frac{1}{4}$

6) $-\frac{5}{3}$

7) $\frac{7}{4}$

8) 4

9) $\frac{3}{2}$

10) $\frac{3}{4}$

11) -1

12) -1

13) $\frac{1}{21}$

14) 1

15) $\frac{7}{5}$

16) 1

17) 3

18) $\frac{2}{3}$

19) $-\frac{3}{2}$

20) 2

21) 1

22) $\frac{7}{3}$

23) $-\frac{1}{2}$

24) 2

25) 3

26) 1

27) 24.2

28) 190.27

29) $12,432.4

30) $10,792.14

Solving Logarithmic Equations

1) $\{100\}$

2) $\{4\}$

3) $\{\frac{1}{64}\}$

4) $\{20\}$

5) $\{\frac{2}{25}\}$

6) $\{\frac{1}{250}\}$

7) $\{2\}$

8) $\{146\}$

9) $\{\frac{1}{4}\}$

10) No Solution

11) $\{\frac{1}{2}\}$

12) $\{-\frac{1}{3}\}$

13) $\{2\}$

14) No Solution

15) $\{1, -\frac{8}{5}\}$

16) $\{\frac{5}{7}\}$

17) $\{2, -2\}$

18) $\{1\}$

19) $x = \frac{8-8e^4}{4e^4}$

20) $\{-\frac{5}{3}\}$

21) $e^3 + 1$

22) $\{6\}$

23) $\{\frac{8}{3}\}$

24) $\{\frac{46}{17}\}$

25) $x = \frac{16e^5-2}{3}$

26) $x = \frac{9}{2}$

27) $x = \frac{11}{4}$

28) $x = \frac{20}{7}$

29) $e^{\frac{\ln(2)}{2}}$

30) No Solution

31) $x > 2$

32) No Solution

33) No Solution

34) $x = -\frac{3}{14}$

35) $x = \frac{5}{2}$

36) $x = \frac{5e+6}{5}$

Chapter 13 :

Geometry and Solid Figures

Topics that you'll practice in this chapter:

✓ Angles

✓ Pythagorean Relationship

✓ Triangles

✓ Polygons

✓ Trapezoids

✓ Circles

✓ Cubes

✓ Rectangular Prism

✓ Cylinder

✓ Pyramids and Cone

Geometry is the archetype of the beauty of the world.

Johannes Kepler

Angles

✎ **What is the value of x in the following figures?**

1)

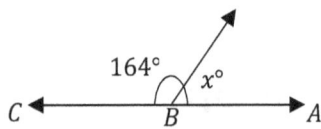

2)

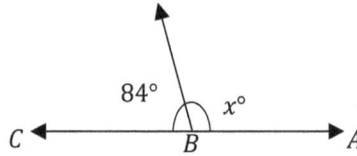

3)

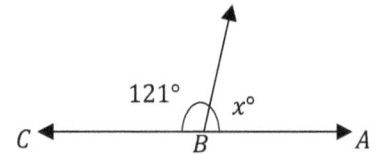

4)

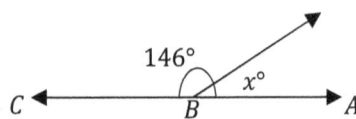

5)

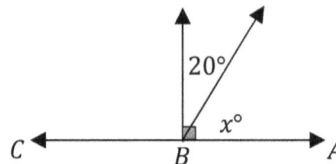

6)

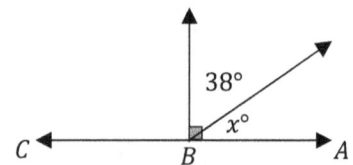

✎ **Calculate.**

7) Two supplement angles have equal measures. What is the measure of each angle? _____

8) The measure of an angle is seven fifth the measure of its supplement. What is the measure of the angle? _____

9) Two angles are complementary and the measure of one angle is 24 less than the other. What is the measure of the smaller angle? _____

10) Two angles are complementary. The measure of one angle is one fifth the measure of the other. What is the measure of the bigger angle? _____

11) Two supplementary angles are given. The measure of one angle is 40° less than the measure of the other. What does the smaller angle measure? _____

Pythagorean Relationship

✎ **Do the following lengths form a right triangle?**

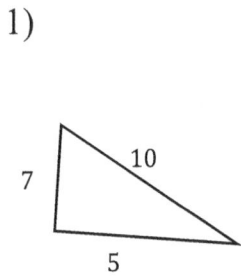

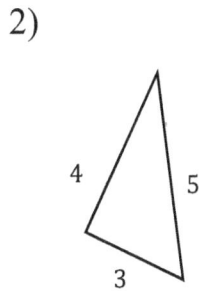

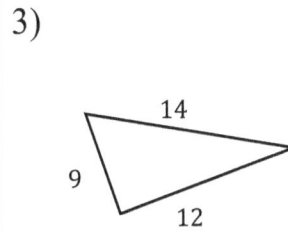

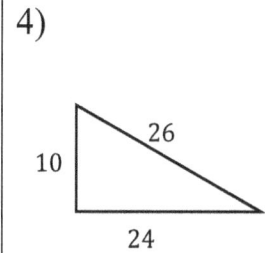

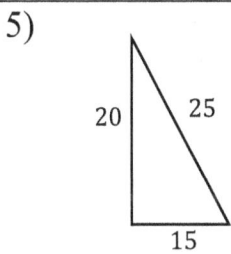

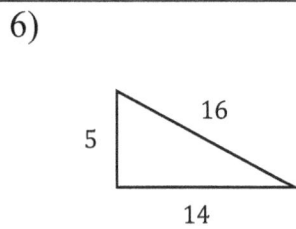

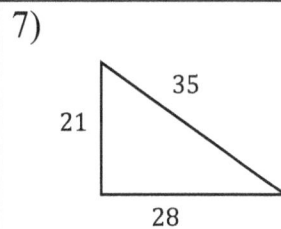

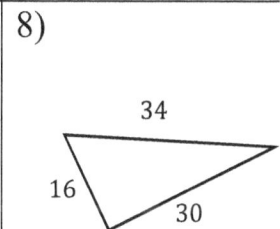

1)

7   10

5

2)

4   5

3

3)

14

9

12

4)

10   26

24

5)

20   25

15

6)

16

5

14

7)

35

21

28

8)

34

16

30

✎ **Find the missing side?**

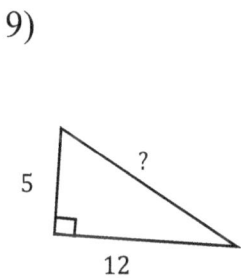

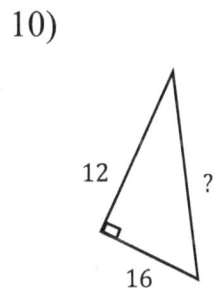

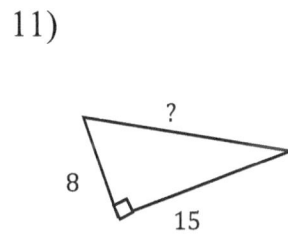

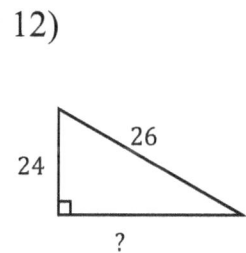

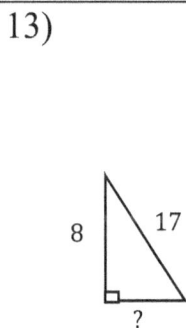

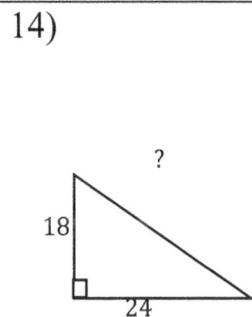

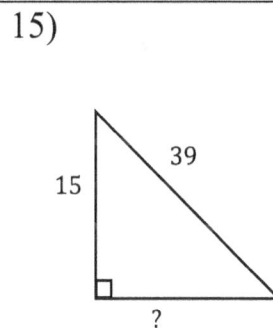

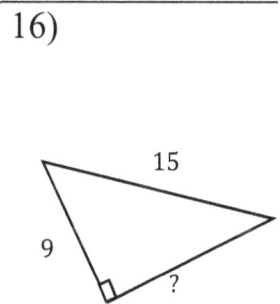

9)

5   ?

12

10)

12   ?

16

11)

?

8

15

12)

26

24

?

13)

8   17

?

14)

?

18

24

15)

39

15

?

16)

15

9

?

Triangles

✎ **Find the measure of the unknown angle in each triangle.**

1)
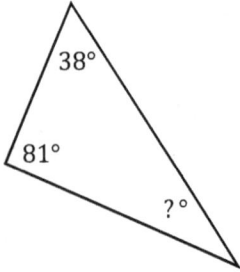
38°
81°
?°

2)
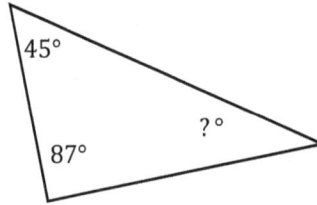
45°
87°
?°

3)
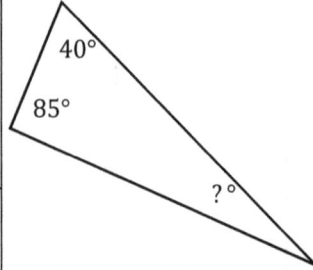
40°
85°
?°

4)
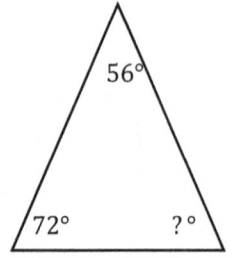
56°
72°
?°

5)
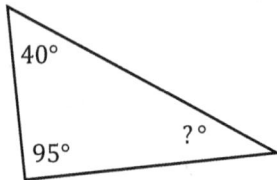
40°
95°
?°

6)
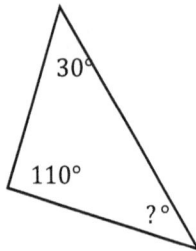
30°
110°
?°

7)
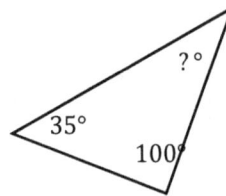
?°
35°
100°

8)
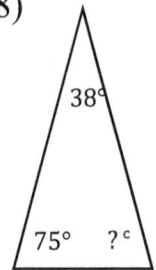
38°
75°
?°

✎ **Find area of each triangle.**

9)
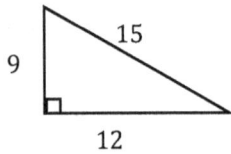
15
9
12

10)
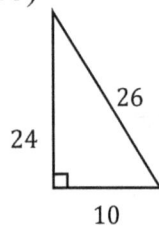
26
24
10

11)
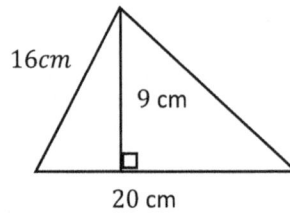
16cm
9 cm
20 cm

12)
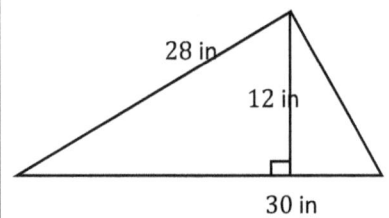
28 in
12 in
30 in

Polygons

✎ **Find the perimeter of each shape.**

1)

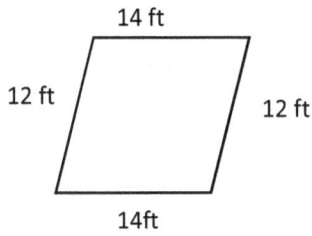

14 ft

12 ft   12 ft

14ft

2)

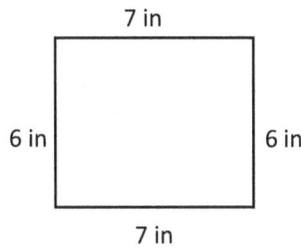

7 in

6 in   6 in

7 in

3)

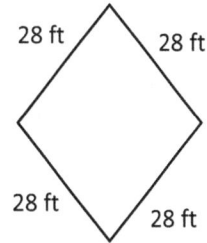

28 ft   28 ft

28 ft   28 ft

4) Square

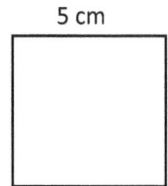

5 cm

5) Regular hexagon

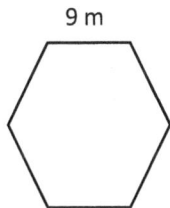

9 m

6)

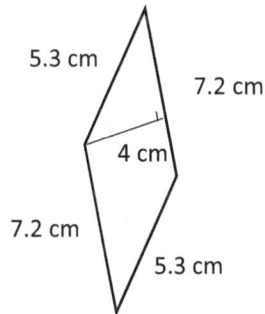

5.3 cm

7.2 cm

4 cm

7.2 cm

5.3 cm

7) Parallelogram

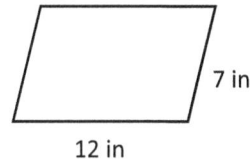

7 in

12 in

8) Square

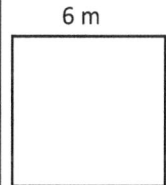

6 m

✎ **Find the area of each shape.**

9) Parallelogram

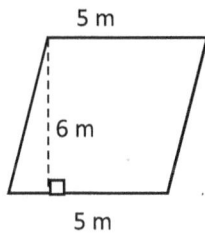

5 m

6 m

5 m

10) Rectangle

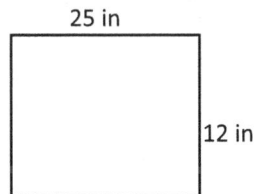

25 in

12 in

11) Rectangle

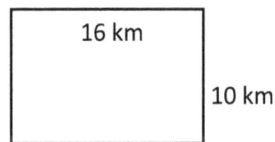

16 km

10 km

12) Square

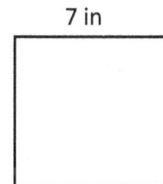

7 in

Trapezoids

✍ **Find the area of each trapezoid.**

1) 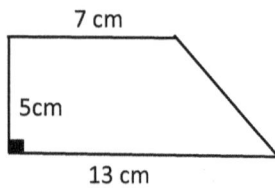 7 cm, 5cm, 13 cm	2) 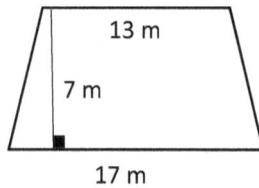 13 m, 7 m, 17 m	3) 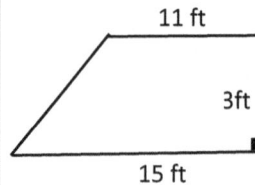 11 ft, 3ft, 15 ft	4) 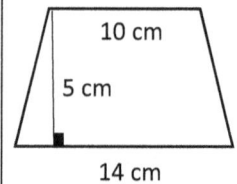 10 cm, 5 cm, 14 cm
5) 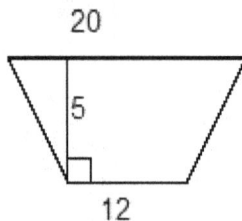 20, 5, 12	6) 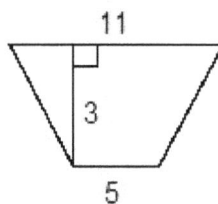 11, 3, 5	7) 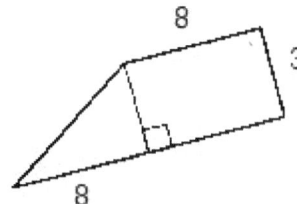 8, 3, 8	8) 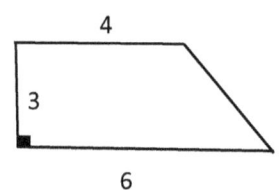 4, 3, 6

✍ **Calculate.**

1) A trapezoid has an area of 45 cm² and its height is 5 cm and one base is 5 cm. What is the other base length? _____

2) If a trapezoid has an area of 99 ft² and the lengths of the bases are 8 ft and 10 ft, find the height? _____

3) If a trapezoid has an area of 126 m² and its height is 14 m and one base is 6 m, find the other base length? _____

4) The area of a trapezoid is 440 ft² and its height is 22 ft. If one base of the trapezoid is 15 ft, what is the other base length?

_____

Circles

✎ **Find the area of each circle.** ($\pi = 3.14$)

1)	2)	3)	4)	5)	6)
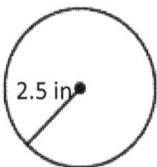	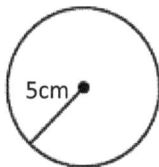	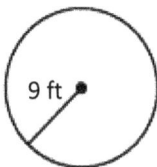	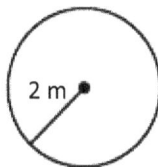	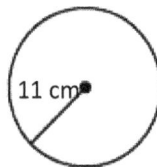	

7)	8)	9)	10)	11)	12)
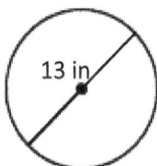	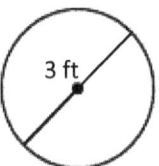	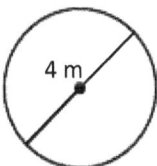	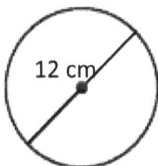		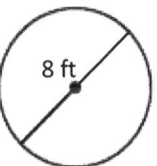

✎ **Complete the table below.** ($\pi = 3.14$)

Circle No.	Radius	Diameter	Circumference	Area
1	1 in	2 in	6.28 in	3.14 in^2
2		10 m		
3				28.26 ft^2
4			47.1 mi	
5		11 km		
6	7 cm			
7		12 ft		
8				314 m^2
9			56.52 in	
10	4.5 ft			

Cubes

✏️ **Find the volume of each cube.**

1)	2)	3)	4)	5)	6)
	2 cm	6 ft	11 m	13 in	7 miles

7)	8	9)	10)	11)	12)
1.2 km	9 cm	2.1 ft	12 mm	0.2 in	0.1 km

✏️ **Find the surface area of each cube.**

13)	14)	15)	16)	17)	18)
	7 m	5 ft	4.5 mm	1.1 km	11 cm

Rectangular Prism

✎ **Find the volume of each Rectangular Prism.**

1)

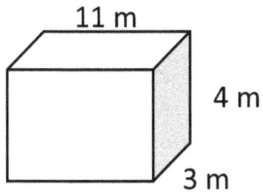

11 m
4 m
3 m

2)

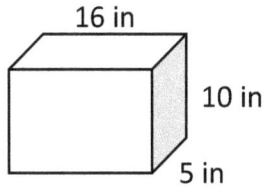

16 in
10 in
5 in

3)

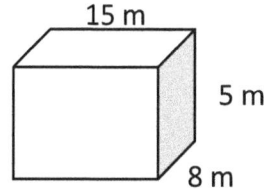

15 m
5 m
8 m

4)

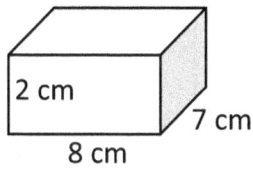

2 cm
7 cm
8 cm

5)

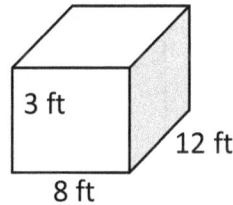

3 ft
12 ft
8 ft

6)

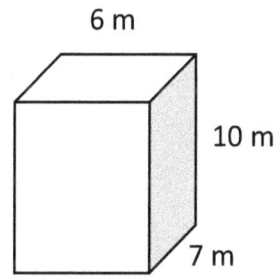

6 m
10 m
7 m

✎ **Find the surface area of each Rectangular Prism.**

7)

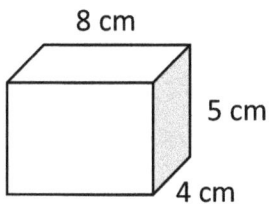

8 cm
5 cm
4 cm

8)

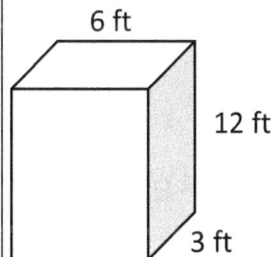

6 ft
12 ft
3 ft

9)

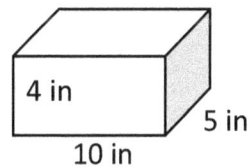

4 in
5 in
10 in

10)

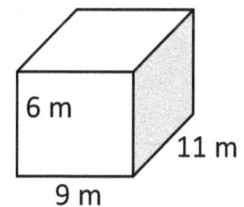

6 m
11 m
9 m

Cylinder

✎ **Find the volume of each Cylinder. Round your answer to the nearest tenth.** ($\pi = 3.14$)

1)

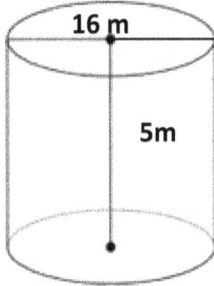

16 m

5m

2)

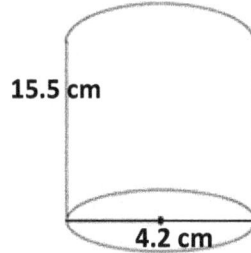

15.5 cm

4.2 cm

3)

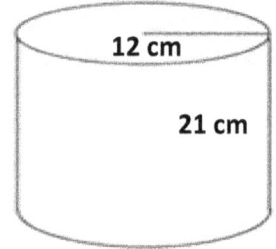

12 cm

21 cm

4)

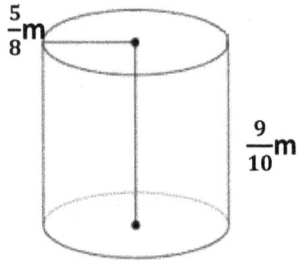

$\frac{5}{8}$ m

$\frac{9}{10}$ m

5)

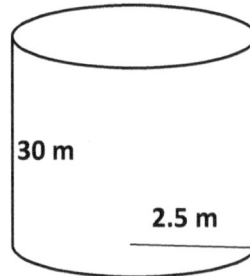

30 m

2.5 m

6)

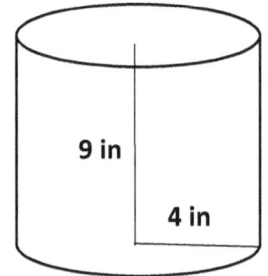

9 in

4 in

✎ **Find the surface area of each Cylinder.** ($\pi = 3.14$)

7)

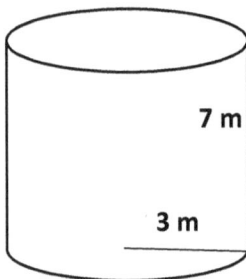

7 m

3 m

8)

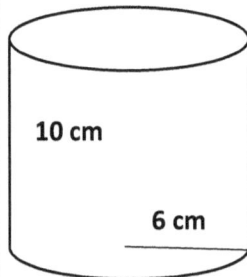

10 cm

6 cm

9)

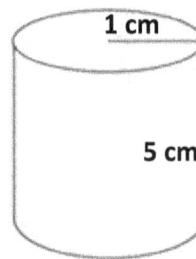

1 cm

5 cm

10)

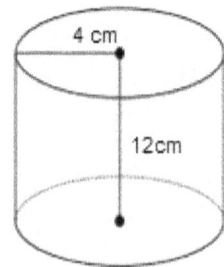

4 cm

12cm

Pyramids and Cone

✎ **Find the volume of each Pyramid and Cone.** ($\pi = 3.14$)

1)	2)	3)
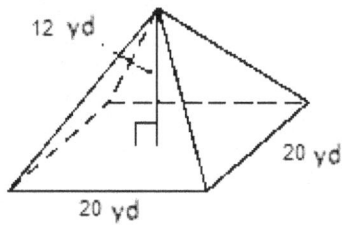	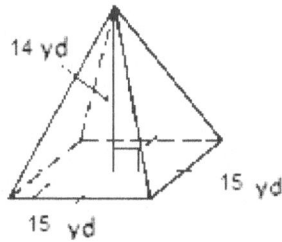	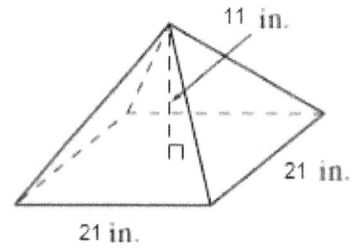

4)	5)	6)
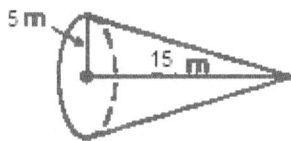	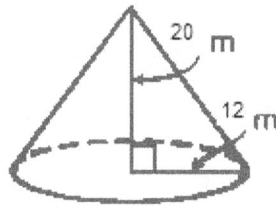	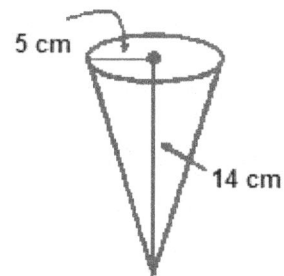

✎ **Find the surface area of each Pyramid and Cone.** ($\pi = 3.14$)

7)	8)	9)	10)
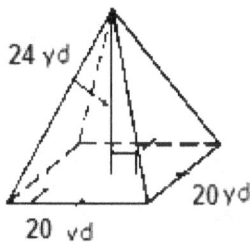	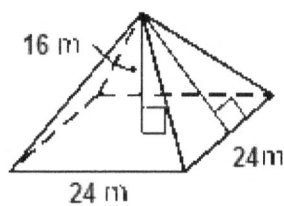	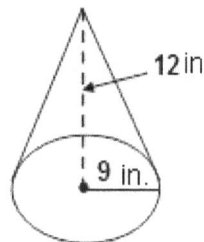	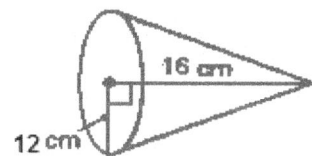

Answers of Worksheets

Angles

1) 16°   4) 34°   7) 90°   10) 75°

2) 96°   5) 70°   8) 75°   11) 70°

3) 59°   6) 52°   9) 33°

Pythagorean Relationship

1) No   5) Yes   9) 13   13) 15

2) Yes   6) No   10) 20   14) 30

3) No   7) Yes   11) 17   15) 36

4) Yes   8) Yes   12) 10   16) 12

Triangles

1) 60°   5) 45°   9) 54 $square\ unites$

2) 48°   6) 40°   10) 120 $square\ unites$

3) 55°   7) 45°   11) 90 $square\ unites$

4) 52°   8) 67°   12) 180 $square\ unites$

Polygons

1) 52 ft 5) 54 m 9) 30 m^2

2) 26 in 6) 25 cm 10) 300 in^2

3) 112 ft 7) 38 in 11) 160 km^2

4) 20 cm 8) 24 m 12) 49 in^2

Trapezoids

1) 50 cm^2 4) 60 cm^2 7) 36

2) 105 m^2 5) 80   8) 15

3) 39 ft^2 6) 24

Calculate

1) 13 cm 2) 11 ft 3) 12 m 4) 25 ft

Circles

1) 19.63 in^2 5) 379.94 cm^2 9) 12.56 m^2

2) 78.5 cm^2 6) 314 $miles^2$ 10) 113.04 cm^2

3) 254.34 ft^2 7) 132.67 in^2 11) 38.47 $miles^2$

4) 12.56 m^2 8) 7.07 ft^2 12) 50.24 ft^2

Circle No.	Radius	Diameter	Circumference	Area
1	1 in	2 in	6.28 in	3.14 in^2
2	5 m	10 m	31.4 m	78.5 m^2
3	3 ft	6 ft	18.84 ft	28.26 ft^2
4	7.5 miles	15 mi	47.1 mi	176.63 mi^2
5	5.5 km	11 km	34.54 km	94.99 km^2
6	7 cm	14 cm	43.96 cm	153.86 cm^2
7	6 ft	12 ft	37.68 feet	113.04 ft^2
8	10 m	20 m	62.8 m	314 m^2
9	9 in	18 in	56.52 in	254.34 in^2
10	4.5 ft	9 ft	28.26 ft	63.585 ft^2

Cubes

1) 12
2) 8 cm^3
3) 216 ft^3
4) 1,331 m^3
5) 2,197 in^3

6) 343 $miles^3$
7) 1.728 km^3
8) 729 cm^3
9) 9.261 ft^3
10) 1,728 mm^3

11) 0.008 in^3
12) 0.001 km^3
13) 27
14) 294 m^2
15) 150 ft^2

16) 121.5 mm^2
17) 7.26 km^2
18) 726 cm^2

Rectangular Prism

1) 132 m^3
2) 800 in^3
3) 600 m^3

4) 112 cm^3
5) 288 ft^3
6) 420 m^3

7) 184 cm^2
8) 252 ft^2
9) 220 in^2

10) 438 m^2

Cylinder

1) 1,004.8 m^3
2) 214.6 cm^3
3) 9,495.4 cm^3

4) 1.1 m^3
5) 588.8 m^3
6) 452.2 in^3

7) 188.4 m^2
8) 602.9 cm^2
9) 37.7 cm^2

10) 401.9 m^2

Pyramids and Cone

1) 1,600 yd^3
2) 1,050 yd^3
3) 1,617 in^3

4) 392.5 m^3
5) 3,014.4 m^3
6) 366.33 cm^3

7) 1,440 yd^2
8) 1,536 m^2
9) 678.24 in^2

10) 1,205.76 cm^2

Chapter 14 :

Trigonometric Functions

Topics that you'll practice in this chapter:

✓ Trig ratios of General Angles

✓ Sketch Each Angle in Standard Position

✓ Finding Co–Terminal Angles and Reference Angles

✓ Angles in Radians

✓ Angles in Degrees

✓ Evaluating Each Trigonometric Expression

✓ Missing Sides and Angles of a Right Triangle

✓ Arc Length and Sector Area

Mathematics is like checkers in being suitable for the young, not too difficult, amusing, and without peril to the state. — *Plato*

Trig ratios of General Angles

✎ **Evaluate.**

1) $\sin 135° =$ _____

2) $\sin 300° =$ _____

3) $\cos -225° =$ _____

4) $\cos 270° =$ _____

5) $\sin 450° =$ _____

6) $\sin -330° =$ _____

7) $\tan 60° =$ _____

8) $\cot 180° =$ _____

9) $\tan 240° =$ _____

10) $\cot 90° =$ _____

11) $\sec 180° =$ _____

12) $\csc 90° =$ _____

13) $\cot -270° =$ _____

14) $\sec 360° =$ _____

15) $\cos -45° =$ _____

16) $\sec 120° =$ _____

17) $\csc 360° =$ _____

18) $\cot -45° =$ _____

✎ **Find the exact value of each trigonometric function. Some may be undefined.**

19) $\sec 2\pi =$ _____

20) $\tan -\dfrac{5\pi}{2} =$ _____

21) $\cos \dfrac{11\pi}{2} =$ _____

22) $\cot \dfrac{9\pi}{4} =$ _____

23) $\sec -6\pi =$ _____

24) $\sec \dfrac{\pi}{4} =$ _____

25) $\csc \dfrac{8\pi}{3} =$ _____

26) $\cot \dfrac{10\pi}{3} =$ _____

27) $\csc -\dfrac{\pi}{2} =$ _____

28) $\cot \dfrac{2\pi}{3} =$ _____

Sketch Each Angle in Standard Position

✎ **Draw each angle with the given measure in standard position.**

1) $-570°$

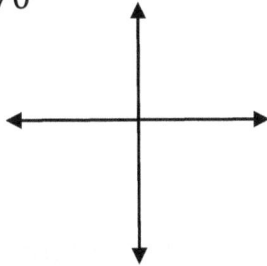

4) $-690°$

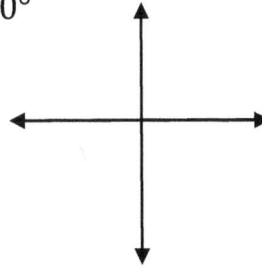

2) $750°$

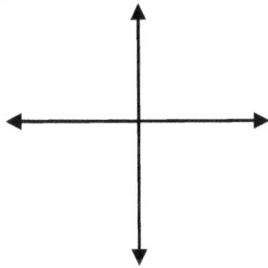

5) $\frac{13\pi}{6}$

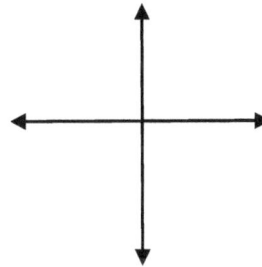

3) $1{,}110°$

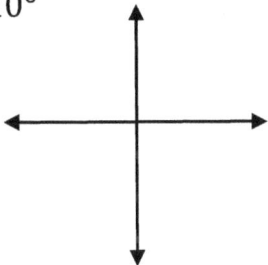

6) $-\frac{11\pi}{6}$

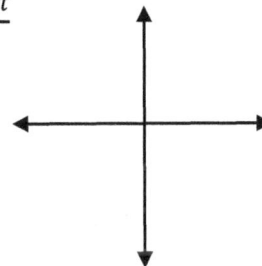

Finding Co-terminal Angles and Reference Angles

✍ **Find a conterminal angle between 0° and 360° for each angle provided.**

1) $-315° =$

3) $-225° =$

2) $-210° =$

4) $-540° =$

✍ **Find a conterminal angle between 0 and 2π for each given angle.**

5) $\dfrac{18\pi}{5} =$

7) $-\dfrac{13\pi}{4} =$

6) $-\dfrac{19\pi}{6} =$

8) $\dfrac{14\pi}{3} =$

✍ **Find the reference angle of each angle.**

9)

10)

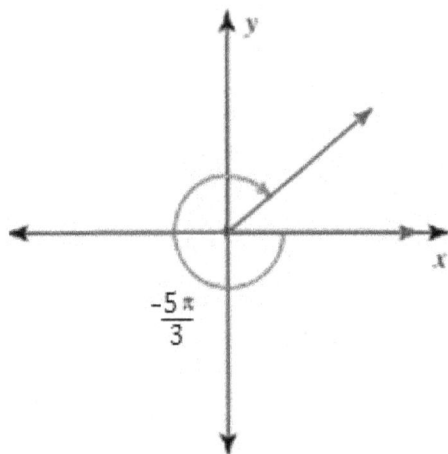

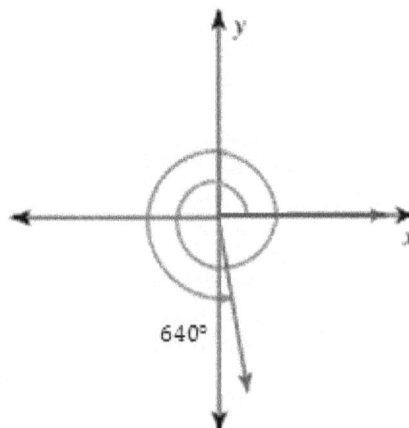

Angles and Angle Measure

✎ **Convert each degree measure into radians.**

1) $216° = $ ____

2) $660° = $ ____

3) $420° = $ ____

4) $220° = $ ____

5) $210° = $ ____

6) $270° = $ ____

7) $-300° = $ ____

8) $810° = $ ____

9) $330° = $ ____

10) $140° = $ ____

11) $480° = $ ____

12) $405° = $ ____

13) $-450° = $ ____

14) $-126° = $ ____

15) $-675° = $ ____

16) $150° = $ ____

17) $-468° = $ ____

18) $340° = $ ____

19) $-440° = $ ____

20) $342° = $ ____

21) $230° = $ ____

✎ **Convert each radian measure into degrees.**

22) $\frac{\pi}{10} = $

23) $\frac{5\pi}{12} = $

24) $\frac{7\pi}{3} = $

25) $\frac{3\pi}{20} = $

26) $-\frac{6\pi}{5} = $

27) $\frac{11\pi}{18} = $

28) $-\frac{14\pi}{5} = $

29) $\frac{5\pi}{18} = $

30) $\frac{7\pi}{36} = $

31) $\frac{17\pi}{18} = $

32) $-\frac{13\pi}{30} = $

33) $\frac{7\pi}{9} = $

34) $-\frac{19\pi}{18} = $

35) $\frac{7\pi}{60} = $

36) $-\frac{3\pi}{10} = $

37) $\frac{11\pi}{30} = $

38) $-\frac{2\pi}{9} = $

39) $-\frac{7\pi}{10} = $

Evaluating Trigonometric Functions

✎ **Find the exact value of each trigonometric function.**

1) $\cos 780° = $ _____

2) $\tan \dfrac{5\pi}{3} = $ _____

3) $\tan -\dfrac{\pi}{6} = $ _____

4) $\cot -\dfrac{9\pi}{4} = $ _____

5) $\cos -\dfrac{7\pi}{6} = $ _____

6) $\cos 135° = $ _____

7) $\sin 240° = $ _____

8) $\tan 330° = $ _____

9) $\cot 420° = $ _____

10) $\tan -495° = $ _____

11) $\cot 315° = $ _____

12) $\sin -240° = $ _____

13) $\cot 225° = $ _____

✎ **Use the given point on the terminal side of angle θ to find the value of the trigonometric function indicated.**

14) $\sin\theta,\ (-6,8)$

15) $\cos\theta,\ (-6,8)$

16) $\sec\theta,\ (3,\ 5)$

17) $\cos\theta,\ (10,24)$

18) $\sin\theta,\ (6,-6)$

19) $\tan\theta,\ (-2,-\sqrt{12})$

Missing Sides and Angles of a Right Triangle

✒ Find the value of each trigonometric ratio as fractions in their simplest form.

1) $\cot x$

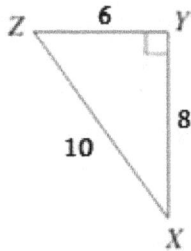

2) $\cos A$

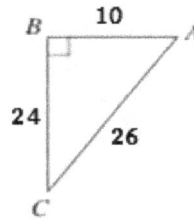

✒ Find the missing sides. Round answers to the nearest tenth.

3)

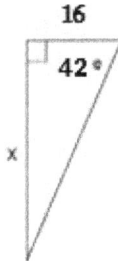

4)

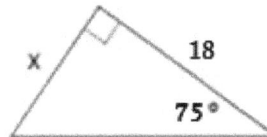

5)

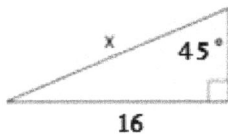

6)

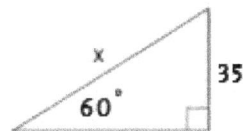

Arc Length and Sector Area

✎ **Find the length of each arc. Round your answers to the nearest tenth.**

$(\pi = 3.14)$

1) $r = 28$ cm, $\theta = 30°$

3) $r = 22$ ft, $\theta = 50°$

2) $r = 14$ ft, $\theta = 95°$

4) $r = 16\,m$, $\theta = 85°$

✎ **Find area of each sector. Do *not* round. Round your answers to the nearest tenth. $(\pi = 3.14)$**

5)

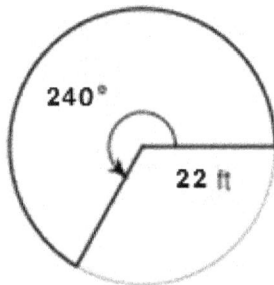

240°
22 ft

7)

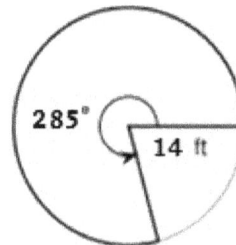

285°
14 ft

6)

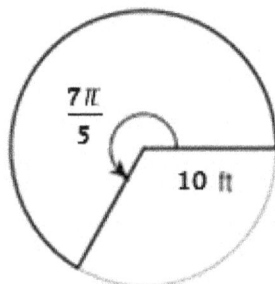

$\dfrac{7\pi}{5}$
10 ft

8)

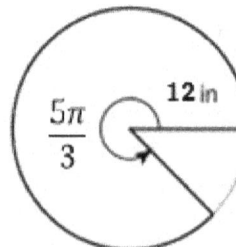

$\dfrac{5\pi}{3}$
12 in

Answers of Worksheets

Trig Ratios of General Angles

1) $\frac{\sqrt{2}}{2}$

2) $-\frac{\sqrt{3}}{2}$

3) $-\frac{\sqrt{2}}{2}$

4) 0

5) 1

6) $\frac{1}{2}$

7) $\sqrt{3}$

8) Undefined

9) $\sqrt{3}$

10) 0

11) -1

12) 1

13) 0

14) 1

15) $\frac{\sqrt{2}}{2}$

16) -2

17) Undefined

18) -1

19) 1

20) Undefined

21) 0

22) 1

23) 1

24) $\sqrt{2}$

25) $\frac{2\sqrt{3}}{3}$

26) $\frac{\sqrt{3}}{3}$

27) -1

28) $-\frac{\sqrt{3}}{3}$

Sketch Each Angle in Standard Position

1) $-570°$

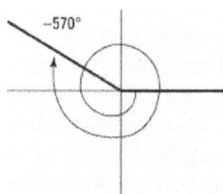

2) $750°$

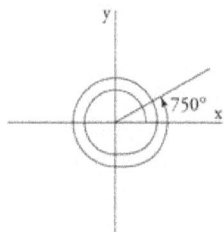

3) $1,110°$

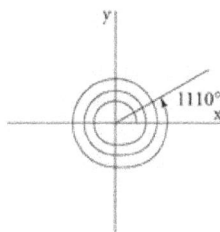

4) $-690°$

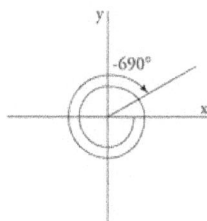

5) $\frac{13\pi}{6} = 390°$

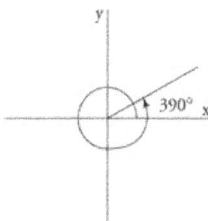

6) $-\frac{11\pi}{6} = -330°$

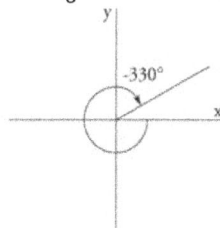

Finding Co–Terminal Angles and Reference Angles

1) $45°$

2) $150°$

3) $135°$

4) $180°$

5) $\frac{4\pi}{5}$

6) $\frac{5\pi}{6}$

7) $\frac{3\pi}{4}$

8) $\frac{2\pi}{3}$

9) $\frac{\pi}{3}$

10) $80°$

Angles and Angle Measure

1) $\frac{6\pi}{5}$

2) $\frac{11\pi}{3}$

3) $\frac{7\pi}{3}$

4) $\frac{11\pi}{9}$

5) $\frac{7\pi}{6}$

6) $\frac{3\pi}{2}$

7) $-\frac{5\pi}{3}$

8) $\frac{9\pi}{2}$

9) $\frac{11\pi}{6}$

10) $\frac{7\pi}{9}$

11) $\frac{8\pi}{3}$

12) $\frac{9\pi}{4}$

13) $-\frac{5}{2}\pi$

14) $-\frac{7\pi}{10}$

15) $-\frac{15\pi}{4}$

16) $\frac{5\pi}{6}$

17) $-\frac{13\pi}{5}$

18) $\frac{17\pi}{9}$

19) $-\frac{22\pi}{9}$

20) $\frac{19\pi}{10}$

21) $\frac{23\pi}{18}$

22) $18°$

23) $75°$

24) $420°$

25) $27°$

26) $-216°$

27) $110°$

28) $-504°$

29) $50°$

30) $35°$

31) $170°$

32) $-78°$

33) $140°$

34) $-190°$

35) $21°$

36) $-54°$

37) $66°$

38) $-40°$

39) $-126°$

Evaluating Each Trigonometric Functions

1) $\frac{1}{2}$

2) $-\sqrt{3}$

3) $-\frac{\sqrt{3}}{3}$

4) -1

5) $-\frac{\sqrt{3}}{2}$

6) $-\frac{\sqrt{2}}{2}$

7) $-\frac{\sqrt{3}}{2}$

8) $-\frac{\sqrt{3}}{3}$

9) $\frac{\sqrt{3}}{3}$

10) 1

11) -1

12) $\frac{\sqrt{3}}{2}$

13) 1

14) 0.8

15) -0.6

16) $\frac{\sqrt{34}}{5}$

17) $\frac{5}{13}$

18) $-\frac{\sqrt{2}}{2}$

19) $\sqrt{3}$

Missing Sides and Angles of a Right Triangle

1) $\frac{4}{3}$

2) $\frac{5}{13}$

3) 14.4

4) 67.2

5) 22.6

6) 40.4

Arc Length and Sector Area

1) $14.7\ cm$

2) $23.2\ ft$

3) $19.2\ ft$

4) $23.7m$

5) $1,013.7\ ft^2$

6) $220\ in^2$

7) $487.5\ ft^2$

8) $377\ in^2$

Chapter 15 :

Statistics and Probability

Topics that you'll practice in this chapter:

- ✓ Mean and Median
- ✓ Mode and Range
- ✓ Histograms
- ✓ Stem–and–Leaf Plot
- ✓ Pie Graph
- ✓ Probability Problems
- ✓ Factorials
- ✓ Combinations and Permutation

"The book of nature is written in the language of Mathematic."

- Galileo.

Mean and Median

🖎 Find Mean and Median of the Given Data.

1) 8, 7, 14, 4, 8

2) 14, 8, 25, 19, 16, 33, 11

3) 23, 18, 15, 12, 17

4) 34, 14, 10, 15, 6, 11

5) 10, 19, 6, 8, 32, 20, 17

6) 17, 26, 39, 69, 20, 6

7) 40, 38, 18, 11, 9, 2, 7, 32, 41

8) 24, 21, 31, 12, 33, 32, 22

9) 16, 14, 20, 41, 15, 20, 38, 4

10) 20, 20, 30, 18, 6, 28, 12, 46

11) 12, 7, 10, 11, 16, 22

12) 10, 29, 27, 12, 2, 15, 10, 3

🖎 Calculate.

13) In a javelin throw competition, five athletics score 56, 34, 62, 23 and 19 meters. What are their Mean and Median? _____

14) Eva went to shop and bought 8 apples, 14 peaches, 6 bananas, 4 pineapples and 12 melons. What are the Mean and Median of her purchase? _____

15) Bob has 17 black pen, 19 red pen, 14 green pens, 20 blue pens and 5 boxes of yellow pens. If the Mean and Median are 19 respectively, what is the number of yellow pens in each box? _____

Mode and Range

✎ **Find Mode and Rage of the Given Data.**

1) 4, 3, 7, 3, 3, 4

    Mode: _____   Range: _____

2) 18, 18, 24, 26, 18, 8, 14, 22

    Mode: _____   Range: _____

3) 8, 8, 8, 16, 19, 22, 20, 9, 13

    Mode: _____   Range: _____

4) 24, 24, 14, 28, 20, 18, 20, 24

    Mode: _____   Range: _____

5) 6, 21, 27, 24, 27, 27

    Mode: _____   Range: _____

6) 21, 8, 8, 7, 8, 12, 10, 22, 18, 13

    Mode: _____   Range: _____

7) 7, 4, 4, 6, 13, 13, 13, 0, 2, 2

    Mode: _____   Range: _____

8) 5, 8, 5, 14, 12, 14, 3, 5, 18

    Mode: _____   Range: _____

9) 7, 7, 7, 12, 7, 3, 8, 16, 3, 17

    Mode: _____   Range: _____

10) 15, 15, 19, 16, 4, 16, 10, 15

    Mode: _____   Range: _____

11) 6, 6, 5, 6, 42, 13, 19, 2

    Mode: _____   Range: _____

12) 8, 8, 9, 8, 9, 4, 34, 22

    Mode: _____   Range: _____

✎ **Calculate.**

13) A stationery sold 12 pencils, 56 red pens, 24 blue pens, 20 notebooks, 12 erasers, 21 rulers and 11 color pencils. What are the Mode and Range for the stationery sells?

           Mode: _____   Range: _____

14) In an English test, eight students score 10, 15, 15, 18 18, 16, 15 and 15. What are their Mode and Range? _____

15) What is the range of the first 6 even numbers greater than 8?

    _____

Times Series

🖎 **Use the following Graph to complete the table.**

Day	Distance (km)
1	
2	

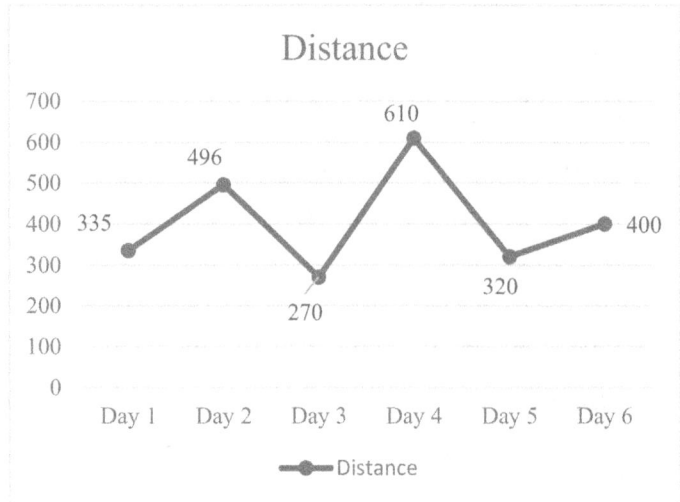

The following table shows the number of births in the US from 2007 to 2012 (in millions).

Year	Number of births (in millions)
2007	4.15
2008	3.70
2009	3.45
2010	3.20
2011	1.75
2012	2.98

Draw a Time Series for the table.

Stem–and–Leaf Plot

✎ **Make stem ad leaf plots for the given data.**

1) 24, 26, 29, 20, 53, 27, 51, 55, 36, 21, 37, 30

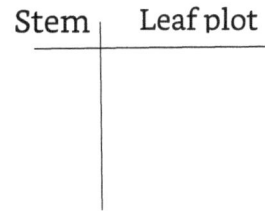

Stem | Leaf plot

2) 11, 59, 66, 14, 18, 19, 59, 65, 69, 61, 68, 65

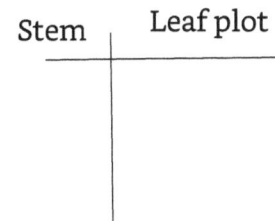

Stem | Leaf plot

3) 121, 55, 66, 54, 112, 128, 63, 125, 59, 123, 68, 119

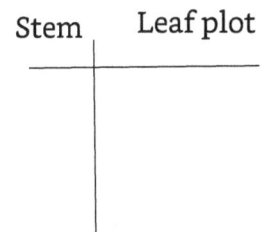

Stem | Leaf plot

4) 51, 32, 100, 56, 84, 36, 107, 56, 85, 39, 56, 106, 89

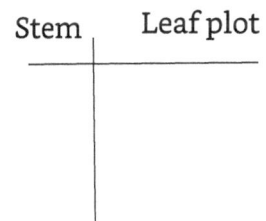

Stem | Leaf plot

5) 33, 89, 19, 87, 81, 16, 11, 30, 86, 35, 17, 35, 13

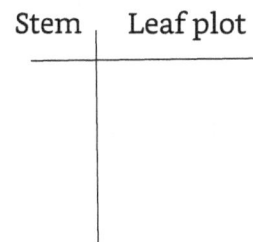

Stem | Leaf plot

6) 60, 92, 22, 25, 67, 93, 95, 62, 21, 64, 98, 29

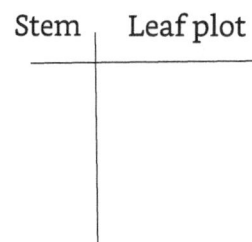

Stem | Leaf plot

Pie Graph

The circle graph below shows all Robert's expenses for last month. Robert spent $140 on his hobbies last month.

Answer following questions based on the Pie graph.

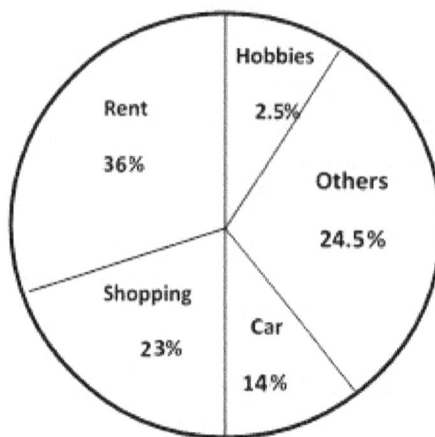

1) How much was Robert's total expenses last month? _____

2) How much did Robert spend on his car last month? _____

3) How much did Robert spend for shopping last month? _____

4) How much did Robert spend on his rent last month? _____

5) What fraction is Robert's expenses for his rent and car out of his total

   expenses last month? _____

Probability Problems

✎ **Calculate.**

1) A number is chosen at random from 1 to 10. Find the probability of selecting number 6 or smaller numbers. _____

2) Bag A contains 18 red marbles and 6 green marbles. Bag B contains 16 black marbles and 8 orange marbles. What is the probability of selecting a green marble at random from bag A? What is the probability of selecting a black marble at random from Bag B? _____

3) A number is chosen at random from 1 to 20. What is the probability of selecting multiples of 4? _____

4) A card is chosen from a well-shuffled deck of 52 cards. What is the probability that the card will be a queen? _____

5) A number is chosen at random from 1 to 15. What is the probability of selecting a multiple of 3 or 5? _____

A spinner numbered 1–8, is spun once. What is the probability of spinning …?

6) an Odd number? _____   7) a multiple of 2? _____

8) a multiple of 5? _____   9) number 10? _____

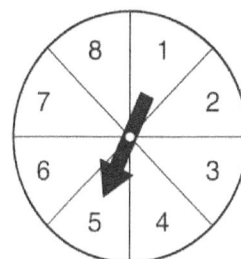

Factorials

✎ **Determine the value for each expression.**

1) $4! + 0! =$

2) $2! + 5! =$

3) $(2!)^2 =$

4) $5! - 3! =$

5) $6! - 3! + 10 =$

6) $3! \times 4 - 15 =$

7) $(2! + 3!)^2 =$

8) $(4! - 3!)^2 =$

9) $(3!\,0!)^2 - 10 =$

10) $\dfrac{10!}{8!} =$

11) $\dfrac{6!}{4!} =$

12) $\dfrac{6!}{5!} =$

13) $\dfrac{15!}{13!} =$

14) $\dfrac{n!}{(n-3)!} =$

15) $\dfrac{(n+2)!}{n!} =$

16) $\dfrac{(2+2!)^3}{2!} =$

17) $\dfrac{5(n+2)!}{(n+1)!} =$

18) $\dfrac{22!}{20!4!} =$

19) $\dfrac{13!}{11!3!} =$

20) $\dfrac{9 \times 210!}{3(7 \times 30)!} =$

21) $\dfrac{32!}{31!2!} =$

22) $\dfrac{11!12!}{10!13!} =$

23) $\dfrac{16!15!}{14!14!} =$

24) $\dfrac{(5 \times 3)!}{0!14!} =$

25) $\dfrac{4!(5n-2)!}{(5n)!} =$

26) $\dfrac{4n(4n+7)!}{(4n+8)!} =$

27) $\dfrac{(n-2)!(n+1)}{(n+2)!} =$

Combinations and Permutations

✍ **Calculate the value of each.**

1) $6! =$ ____

2) $2! \times 5! =$ ____

3) $3 \times 4! =$ ____

4) $5! + 3! =$ ____

5) $7! =$ ____

6) $4! =$ ____

7) $3! + 3! =$ ____

8) $7! - 5! =$ ____

✍ **Find the answer for each word problems.**

9) Susan is baking cookies. She uses sugar, butter, Vanilla, eggs and flour. How many different orders of ingredients can she try? _____

10) Albert is planning for his vacation. He wants to go to museum, watch a movie, go to the beach, play the game and play football. How many ways of ordering are there for him? _____

11) How many 4-digit numbers can be named using the digits 3, 4, 5, and 6 without repetition? _____

12) In how many ways can 5 boys be arranged in a straight line? _____

13) In how many ways can 6 athletes be arranged in a straight line? _____

14) A professor is going to arrange her 7 students in a straight line. In how many ways can she do this? _____

15) How many code symbols can be formed with the letters for the word GAMES? _____

16) In how many ways a team of 7 basketball players can choose a captain and co-captain? _____

Answers of Worksheets

Mean and Median

1) Mean: 8.2, Median: 8
2) Mean: 18, Median: 16
3) Mean: 17, Median: 17
4) Mean: 15, Median: 12.5
5) Mean: 16, Median: 17
6) Mean: 29.5, Median: 23
7) Mean: 22, Median: 18
8) Mean: 25, Median: 24
9) Mean: 21, Median: 18
10) Mean: 22.5, Median: 20
11) Mean: 13, Median: 11.5
12) Mean: 13.5, Median: 11
13) Mean: 38.8, Median: 34
14) Mean: 8.8, Median: 8
15) 5

Mode and Range

1) Mode: 3, Range: 4
2) Mode: 18, Range: 18
3) Mode: 8, Range: 14
4) Mode: 24, Range: 14
5) Mode: 27, Range: 21
6) Mode: 8, Range: 15
7) Mode: 13, Range: 13
8) Mode: 5, Range: 15
9) Mode: 7, Range: 14
10) Mode: 15, Range: 15
11) Mode: 6, Range: 40
12) Mode: 8, Range: 30
13) Mode: 12, Range: 45
14) Mode: 15, Range: 8
15) 10

Time series

Day	Distance (km)
1	335
2	496
3	270
4	610
5	320
6	400

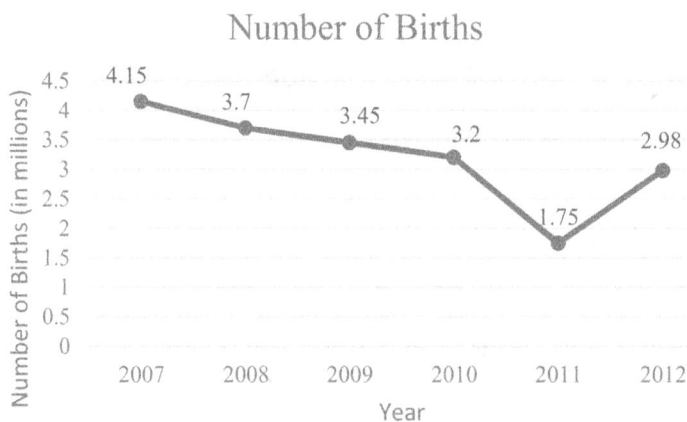

Number of Births

Year	Number of Births (in millions)
2007	4.15
2008	3.7
2009	3.45
2010	3.2
2011	1.75
2012	2.98

Stem–And–Leaf Plot

1)

Stem	leaf
2	0 1 4 6 7 9
3	0 6 7
5	1 3 5

2)

Stem	leaf
1	1 4 8 9
5	9 9
6	1 5 5 6 8 9

3)

Stem	leaf
5	4 5 9
6	3 6 8
11	2 9
12	1 3 5 8

4)

Stem	leaf
3	2 6 9
5	1 6 6 6
8	4 5 9
10	0 6 7

5)

Stem	leaf
1	1 3 6 7 9
3	0 3 5 5
8	1 6 7 9

6)

Stem	leaf
2	2 1 5 9
6	0 2 4 7
9	2 3 5 8

Pie Graph

1) $5,600

2) $784

3) $1,288

4) $2,016

5) $\frac{1}{2}$

Probability Problems

1) $\frac{3}{5}$

2) $\frac{1}{4}, \frac{2}{3}$

3) $\frac{1}{4}$

4) $\frac{1}{13}$

5) $\frac{7}{15}$

6) $\frac{1}{2}$

7) $\frac{1}{2}$

8) $\frac{1}{8}$

9) 0

Factorials

1) 25

2) 122

3) 4

4) 114

5) 724

6) 9

7) 64

8) 324

9) 26

10) 90

11) 30

12) 6

13) 210

14) $n(n-1)(n-2)$

15) $(n+1)(n+2)$

16) 32

17) $5(n+2)$

18) 19.25

19) 26

20) 3

21) 16

22) $\frac{11}{13}$

23) 3,600

24) 15

25) $\frac{24}{5n(5n-1)}$

26) $\frac{n}{(n+2)}$

27) $\frac{1}{n(n-1)(n+2)}$

Combinations and Permutations

1) 720

2) 240

3) 72

4) 126

5) 5,040

6) 24

7) 12

8) 4,920

9) 120

10) 120

11) 24

12) 120

13) 720

14) 5,040

15) 120

16) 42

Chapter 16 :

Next-Generation Accuplacer Test Review

The Next-Generation Accuplacer test is an assessment system for measuring students' readiness for college courses in reading, writing, and mathematics. The test is a multiple–choice format and is used to precisely placing you at the correct level of introductory classes.

The Next-Generation Accuplacer uses the computer–adaptive technology and the questions you see are based on your skill level. Your response to each question drives the difficulty level of the next question.

There are five sub sections on the Accuplacer test:

✓ Arithmetic (20 questions)

✓ Quantitative Reasoning, Algebra, And Statistics (QAS) (20 questions)

✓ Advanced Algebra and Functions (20 questions)

✓ Reading (20 questions)

✓ Writing (25 questions)

Accuplacer does NOT permit the use of personal calculators on the Math portion of placement test. Accuplacer expects students to be able to answer certain questions without the assistance of a calculator. Therefore, they provide an onscreen calculator for students to use on some questions.

In this section, there are two complete Accuplacer Mathematics Tests. Take these tests to see what score you'll be able to receive on a real Accuplacer test.

Time to Test

Time to refine your skill with a practice examination.

Take a practice Accuplacer Math Test to simulate the test day experience. After you've finished, score your test using the answer key.

Before You Start

- You'll need a pencil, a calculator and a timer to take the test.
  - For each question, there are four possible answers. Choose which one is best.
  - It's okay to guess. There is no penalty for wrong answers.
  - After you've finished the test, review the answer key to see where you went wrong.

   The hardest arithmetic to master is that which enables us to count our blessings. ¯Eric Hoffer

Good Luck!

Accuplacer Math Practice Test Answer Sheets

Remove (or photocopy) these answer sheets and use them to complete the practice tests.

Accuplacer Math Practice Test

Section 1		Section 2		Section 3	
1	Ⓐ Ⓑ Ⓒ Ⓓ	1	Ⓐ Ⓑ Ⓒ Ⓓ	1	Ⓐ Ⓑ Ⓒ Ⓓ
2	Ⓐ Ⓑ Ⓒ Ⓓ	2	Ⓐ Ⓑ Ⓒ Ⓓ	2	Ⓐ Ⓑ Ⓒ Ⓓ
3	Ⓐ Ⓑ Ⓒ Ⓓ	3	Ⓐ Ⓑ Ⓒ Ⓓ	3	Ⓐ Ⓑ Ⓒ Ⓓ
4	Ⓐ Ⓑ Ⓒ Ⓓ	4	Ⓐ Ⓑ Ⓒ Ⓓ	4	Ⓐ Ⓑ Ⓒ Ⓓ
5	Ⓐ Ⓑ Ⓒ Ⓓ	5	Ⓐ Ⓑ Ⓒ Ⓓ	5	Ⓐ Ⓑ Ⓒ Ⓓ
6	Ⓐ Ⓑ Ⓒ Ⓓ	6	Ⓐ Ⓑ Ⓒ Ⓓ	6	Ⓐ Ⓑ Ⓒ Ⓓ
7	Ⓐ Ⓑ Ⓒ Ⓓ	7	Ⓐ Ⓑ Ⓒ Ⓓ	7	Ⓐ Ⓑ Ⓒ Ⓓ
8	Ⓐ Ⓑ Ⓒ Ⓓ	8	Ⓐ Ⓑ Ⓒ Ⓓ	8	Ⓐ Ⓑ Ⓒ Ⓓ
9	Ⓐ Ⓑ Ⓒ Ⓓ	9	Ⓐ Ⓑ Ⓒ Ⓓ	9	Ⓐ Ⓑ Ⓒ Ⓓ
10	Ⓐ Ⓑ Ⓒ Ⓓ	10	Ⓐ Ⓑ Ⓒ Ⓓ	10	Ⓐ Ⓑ Ⓒ Ⓓ
11	Ⓐ Ⓑ Ⓒ Ⓓ	11	Ⓐ Ⓑ Ⓒ Ⓓ	11	Ⓐ Ⓑ Ⓒ Ⓓ
12	Ⓐ Ⓑ Ⓒ Ⓓ	12	Ⓐ Ⓑ Ⓒ Ⓓ	12	Ⓐ Ⓑ Ⓒ Ⓓ
13	Ⓐ Ⓑ Ⓒ Ⓓ	13	Ⓐ Ⓑ Ⓒ Ⓓ	13	Ⓐ Ⓑ Ⓒ Ⓓ
14	Ⓐ Ⓑ Ⓒ Ⓓ	14	Ⓐ Ⓑ Ⓒ Ⓓ	14	Ⓐ Ⓑ Ⓒ Ⓓ
15	Ⓐ Ⓑ Ⓒ Ⓓ	15	Ⓐ Ⓑ Ⓒ Ⓓ	15	Ⓐ Ⓑ Ⓒ Ⓓ
16	Ⓐ Ⓑ Ⓒ Ⓓ	16	Ⓐ Ⓑ Ⓒ Ⓓ	16	Ⓐ Ⓑ Ⓒ Ⓓ
17	Ⓐ Ⓑ Ⓒ Ⓓ	17	Ⓐ Ⓑ Ⓒ Ⓓ	17	Ⓐ Ⓑ Ⓒ Ⓓ
18	Ⓐ Ⓑ Ⓒ Ⓓ	18	Ⓐ Ⓑ Ⓒ Ⓓ	18	Ⓐ Ⓑ Ⓒ Ⓓ
19	Ⓐ Ⓑ Ⓒ Ⓓ	19	Ⓐ Ⓑ Ⓒ Ⓓ	19	Ⓐ Ⓑ Ⓒ Ⓓ
20	Ⓐ Ⓑ Ⓒ Ⓓ	20	Ⓐ Ⓑ Ⓒ Ⓓ	20	Ⓐ Ⓑ Ⓒ Ⓓ

Next-Generation Accuplacer Practice Test 1

Section 1

Arithmetic

❖ **20 Questions.**

❖ **Total time for this test: No time limit**.

❖ **You may NOT use a calculator on this Section.**

(On a real Accuplacer test, there is an onscreen calculator to use on some

questions).

Administered *Month Year*

1) Which of the following is closest to 8.12?

   A. 7

   B. 9

   C. 8

   D. 9.1

2) Which of the following is greater than $\frac{14}{5}$?

   A. $\frac{1}{2}$

   B. $\frac{9}{2}$

   C. $\frac{5}{2}$

   D. 1

3) What is the value of 2.85×8.2?

   A. 23.37

   B. 2.337

   C. 233.7

   D. 33.37

4) Which of the following inequalities is true?

   A. $\frac{16}{26} < \frac{7}{13}$

   B. $\frac{5}{9} < \frac{1}{3}$

   C. $\frac{5}{6} < \frac{23}{42}$

   D. $\frac{11}{24} < \frac{7}{8}$

5) 45% of what number is equal to 54?

   A. 1.2

   B. 12

   C. 120

   D. 100

6) The price of a sofa is decreased by 38% to $279. What was its original price?

   A. $106.02

   B. $102.06

   C. $450

   D. $480

7) If 65% of a class are girls, and 20% of girls play tennis, what percent of the class play tennis?

    A. 13%   C. 10%

    B. 14%   D. 87%

8) If 84% of A is 14% of B, then B is what percent of A?

    A. 6%   C. 600%

    B. 60%   D. 0.6%

9) A taxi driver earns $9 per 1-hour work. If he works 10 hours a day and in 1 hour, he uses 1.5-liters petrol with price $3 for 1-liter. How much money does he earn in one day?

    A. $43   C. $54

    B. $42   D. $45

10) The price of a car was $54,000 in 2014, $40,500 in 2016 and $30,375 in 2017. What is the rate of depreciation of the price of car per year?

    A. 15%   C. 30%

    B. 25%   D. 20%

11) Which of the following expressions has the same value as $\frac{24}{7} \times \frac{14}{5}$?

    A. $\frac{8 \times 4}{5}$ C. $\frac{11 \times 4}{5}$

    B. $\frac{12 \times 2}{5}$ D. $\frac{24 \times 2}{5}$

12) A bank is offering 2.5% simple interest on a savings account. If you deposit

$35,000, how much interest will you earn in four years?

    A. $250   C. $3,500

    B. $350   D. $2,500

13) Which of the following lists shows the fractions in order from least to greatest?

$$\frac{4}{19}, \frac{7}{12}, \frac{8}{9}, \frac{5}{16}$$

    A. $\frac{5}{16}, \frac{8}{9}, \frac{7}{12}, \frac{4}{19}$ C. $\frac{4}{19}, \frac{5}{16}, \frac{7}{12}, \frac{8}{9}$

    B. $\frac{8}{9}, \frac{7}{12}, \frac{4}{19}, \frac{5}{16}$ D. $\frac{4}{19}, \frac{7}{12}, \frac{5}{16}, \frac{8}{9}$

14) Sophia purchased a sofa for $472.50. The sofa is regularly priced at $630. What

was the percent discount Sophia received on the sofa?

    A. 20%   C. 30%

    B. 25%   D. 15%

15) A rope weighs 500 grams per meter of length. What is the weight in kilograms

of 21.4 meters of this rope? (1 kilograms = 1,000 grams)

    A. 0.107   C. 10.7

    B. 1.07   D. 107

16) $\frac{19}{6} - \frac{14}{3} = ?$

    A. -0.75 C. 1.5

    B. 0.75   D. -1.5

17) When 61 is divided by 7, the remainder is the same as when 41 is divided by

   A. 11   C. 9

   B. 10   D. 8

18) 24 is What percent of 60?

   A. 60%   C. 40%

   B. 80%   D. 140%

19) How long does a 620–miles trip take moving at 50 miles per hour (mph)?

   A. 12 hours   C. 12 hours and 24 minutes

   B. 12 hours and 4 minutes   D. 12 hours and 40 minutes

20) What is the value of 7.51 + 0.234 + 0.6367?

   A. 8.8307   C. 8.3708

   B. 8.3807   D. 8.3087

STOP

This is the End of Section 1 of Test 1. You may check your work on this Section if you still have time.

Next-Generation Accuplacer Practice Test 1

Section 2

Quantitative Reasoning, Algebra, and Statistics

❖ **20 Questions.**

❖ **Total time for this test: No time limit**.

❖ **You may NOT use a calculator on this Section.**

(On a real Accuplacer test, there is an onscreen calculator to use on some

questions).

Administered *Month Year*

1) Four seventh of 35 is equal to $\frac{5}{9}$ of what number?

    A. 28   C. 35

    B. 36   D. 38

2) A boat sails 72 miles south and then 30 miles east. How far is the boat from its start point?

    A. 87 miles   C. 94 miles

    B. 78 miles   D. 81 miles

3) The ratio of boys and girls in a class is 3:7. If there are 90 students in the class, how many more boys should be enrolled to make the ratio 1:1?

    A. 17   C. 36

    B. 34   D. 27

4) The score of Emma was quarter as that of Ava and the score of Mia was twice that of Ava. If the score of Mia was 80, what is the score of Emma?

    A. 15   C. 5

    B. 10   D. 40

5) An angle is equal to one fifth of its supplement. What is the measure of that angle?

    A. 30   C. 15

    B. 36   D. 20

6) When a number is subtracted from 70 and the difference is divided by that number, the result is 9. What is the value of the number?

A. 8   C. 7

B. 9   D. 5

7) John traveled 150 km in 5 hours and Alice traveled 280 km in 4 hours. What is the ratio of the average speed of John to average speed of Alice?

A. 7: 3   C. 3: 7

B. 5: 8   D. 9: 5

8) A taxi driver earns \$5 per 1-hour work. If he works 12 hours a day and in 1 hour, he uses 2.5-liters petrol with price \$1.2 for 1-liter. How much money does he earn in one day?

A. \$96   C. \$24

B. \$36   D. \$26

9) If the area of trapezoid is 162, what is the perimeter of the trapezoid?

A. 25

B. 52

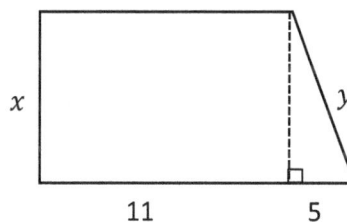

C. 52.50

D. 25.50

10) Right triangle ABC has two legs of lengths 15 cm (AB) and 36 cm (AC). What is the length of the third side (BC)?

   A. 45 cm   C. 35 cm

   B. 42 cm   D. 39 cm

11) The area of a circle is less than 169π. Which of the following can be the circumference of the circle?

   A. 26π C. 30π

   B. 27π D. 34π

12) The width of a box is one sixth of its length. The height of the box is one fourth of its width. If the length of the box is 120 cm, what is the volume of the box?

   A. 9,600 cm^3 C. 15,000 cm^3

   B. 2,400 cm^3 D. 12,000 cm^3

13) How many possible outfit combinations come from six shirts, seven slacks, and five times?

   A. 42   C. 120

   B. 105   D. 210

14) In the xy-plane, the point (2, 2) and (4, 6) are on the line A. Which of the following points could also be on the line A? (Select one or more answer choices)

   A. $(-1, 5)$ C. $(-2, -3)$

   B. $(2, 6)$ D. $(-5, -12)$

15) A bag contains 15 balls: five green, two black, four blue, a brown, a red and two white. If 8 balls are removed from the bag at random, what is the probability that a brown ball has been removed?

A. $\frac{2}{15}$

C. $\frac{8}{15}$

B. $\frac{1}{8}$

D. $\frac{1}{15}$

16) The average of five consecutive numbers is 84. What is the smallest number?

A. 86

C. 82

B. 83

D. 84

17) The average weight of 18 girls in a class is 45 kg and the average weight of 32 boys in the same class is 55 kg. What is the average weight of all the 50 students in that class?

A. 52

C. 52.80

B 51.40

D. 51.80

18) What is the median of these numbers? 10, 24, 12, 18, 37, 29, 23

A.18

C. 12

B. 23

D. 29

19) The surface area of a cylinder is $160\pi \ cm^2$. If its height is 2 cm, what is the radius of the cylinder?

A. 10 cm

C. 6 cm

B. 12 cm

D. 8 cm

20) In 1999, the average worker's income increased $5,000 per year starting from

$36,000 annual salary. Which equation represents income greater than average?

(I = income, x = number of years after 1999)

  A. $I > 5,000\ x + 36,000$

  B. $I > -5,000\ x + 36,000$

  C. $I < -5,000\ x + 36,000$

  D. $I < 5,000\ x - 36,000$

STOP

This is the End of Section 2 of Test 1. You may check your work on this section if you still

have time.

Next-Generation Accuplacer Practice Test 1

Section 3

Advanced Algebra and Functions

❖ **20 Questions.**

❖ **Total time for this test: No time limit**.

❖ **You may use a calculator on this Section.**

(On a real Accuplacer test, there is an onscreen calculator to use on some

questions).

Administered *Month Year*

1) Calculate $f(-2)$ for the following function f.

$$f(x) = 3x^2 - 5x + 2$$

A. 4   C. 0

B. 24   D. 20

2) If $\frac{15}{x} = \frac{20}{x-5}$ what is the value of $\frac{x}{5}$?

A. -5 C. -3

B. -15 D. 3

3) If $x - y = 0$, $2x + 5y = 21$, which of the following ordered pairs (x, y) satisfies both equations?

A. $(2, 3)$ C. $(3, 3)$

B. $(-4, 1)$ D. $(-1, 5)$

4) If $f(x) = 4x + 3(x + 2) - 5$ then $f(3x) = ?$

A. $21x + 1$ C. $x + 21$

B. $21x - 1$ D. $x + 12$

5) A line in the xy-plane passes through origin and has a slope of $\frac{1}{5}$. Which of the following points lies on the line?

A. $(2, 5)$ C. $(10, 2)$

B. $(15, 2)$ D. $(4, 1)$

6) Which of the following is equivalent to $(7n^2 + 9n + 13) - (6n^2 + 5)$?

  A. $9n + 8n^2$ C. $n^2 + 9n + 8$

  B. $n^2 - 5$ D. $9n + 8$

7) If $(ax + 5)(bx + 4) = 21x^2 + cx + 20$ for all values of x and $a + b = 10$,

  what are the two possible values for c?

  A. $54, 59$ C. $43, 47$

  B. $52, 54$ D. $42, 48$

8) If $x \neq -2$ and $x \neq 5$, which of the following is equivalent to $\dfrac{1}{\frac{1}{x-5}+\frac{1}{x+2}}$?

  A. $\dfrac{(x-5)(x+2)}{(x-5)+(x+2)}$ C. $\dfrac{(x+2)(x-5)}{(x+2)-(x+5)}$

  B. $\dfrac{(x+2)+(x-5)}{(x+2)(x-5)}$ D. $\dfrac{(x+2)+(x-5)}{(x+2)-(x-5)}$

9) Which of the following points lies on the line that goes through the points $(1, 5)$

  and $(-1, 7)$?

  A. $(4, 1)$ C. $(9, -1)$

  B. $(2, 3)$ D. $(3, 3)$

10) A function $g(1) = 7$ and $g(4) = 2$. A function $f(7) = 6$ and $f(2) = 14$.

  What is the value of $f(g(4))$?

  A. 14 C. 6

  B. 7 D. 2

$$y > b + x \, , y < x + a$$

11) In the xy-plane, if $(0, 0)$ is a solution to the system of inequalities above, which

of the following relationships between a and b must be true?

A. $a < b$ C. $a = b$

B. $a > b$ D. $a = b - a$

12) What is the area of the following equilateral triangle if the side AB = 8 cm?

A. $6\sqrt{3}\ \text{cm}^2$

B. $16\sqrt{3}\ \text{cm}^2$

C. $8\sqrt{3}\ \text{cm}^2$

D. $12\ \text{cm}^2$

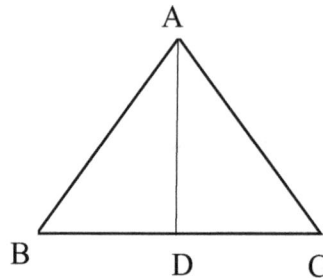

13) A function $g(x)$ satisfies $g(4) = 6$ and $g(9) = 7$. A function $f(x)$ satisfies

$f(6) = 12$ and $f(7) = 26$. What is the value of $f\big(g(9)\big)$?

A. 72   C. 26

B. 182   D. 112

$$x^2 + y^2 - 6x + 12y = 19$$

14) The equation of a circle in the xy −plane is shown above. What is the radius of

the circle?

A. 6   C. 8

B. $\sqrt{28}$ D. $\sqrt{8}$

15) Right triangle ABC is shown below. Which of the following is true for all possible values of angle A and B?

A. $\cos B = \sin A$

B. $\cos B = -\sin A$

C. $\cot A = \cot B$

D. $\cot A = -\cot B$

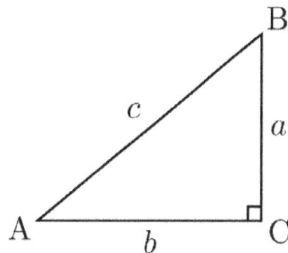

16) John buys a pepper plant that is 8 inches tall. With regular watering the plant grows 4 inches a year. Writing John's plant's height as a function of time, what does the y −intercept represent?

A. The y −intercept represents the rate of grows of the plant which is 8 inches.

B. The y −intercept represents the starting height of 8 inches.

C. The y −intercept represents the rate of growth of plant which is 4 inches per year.

D. There is no y −intercept.

17) Which of the following is an equation of a circle in the xy-plane with center $(0, 4)$ and a radius with endpoint $(\frac{3}{2}, 2)$?

A. $(x + 1)^2 + (y - 4)^2 = \frac{25}{4}$

B. $5x^2 + (y + 4)^2 = \frac{25}{4}$

C. $(x - \frac{3}{2})^2 + (y - 4)^2 = \frac{25}{4}$

D. $x^2 + (y - 4)^2 = \frac{25}{4}$

18) Given a right triangle ΔABC whose $\angle B = 90°$, $\sin C = \frac{5}{13}$, find $\cos A$?

    A. $\frac{7}{13}$ C. $\frac{5}{13}$

    B. $\frac{12}{13}$ D. $\frac{13}{5}$

19) What is the equation of the following graph?

    A. $x^2 + 3x - 10$

    B. $3x^2 - x + 10$

    C. $x^2 - 6x - 10$

    D. $x^2 + 6x + 10$

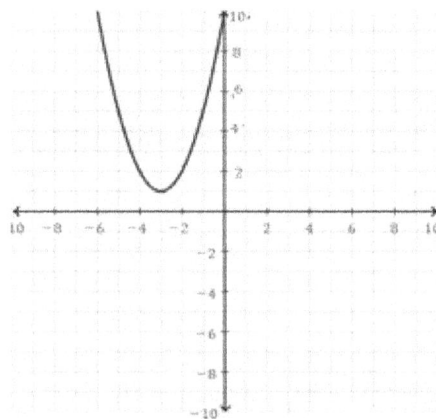

20) In the xy −plane, the line determined by the points $(5, m)$ and $(m, 15)$ passes through the origin. Which of the following could be the value of m?

    A. $\sqrt{15}$ C. $5\sqrt{3}$

    B. 15   D. 3

STOP

This is the End of this Test. You may check your work on this Test if you still have time.

Next-Generation Accuplacer Practice Test 2

Section 1

Arithmetic

❖ **20 Questions.**

❖ **Total time for this test: No time limit**.

❖ **You may NOT use a calculator on this Section.**

(On a real Accuplacer test, there is an onscreen calculator to use on some questions).

Administered *Month Year*

1) What is $53,126 + 7,492$?

    A. 60,618   C. 60,508

    B. 60,816   D. 60,718

2) Bob deposits 15% of $420 into a savings account, what is the amount of his deposit?

    A. $75   C. $36

    B. $57   D. $63

3) What is 4 percent of 450?

    A. 28   C. 18

    B. 10   D. 20

4) In two successive years, the population of a town is increased by 15% and 20%. What percent of the population is increased after two years?

    A. 64%   C. 36%

    B. 62%   D. 38%

5) For what price is 30 percent off the same as $90 off?

    A. $300   C. $150

    B. $450   D. $250

6) What is the remainder when 530 is divided by 7?

    A. 9   C. 5

    B. 6   D. 7

7) Last week 15,000 fans attended a football match. This week four times as many bought tickets, but one eighth of them cancelled their tickets. How many are attending this week?

A. 54,375

C. 52,500

B. 52,000

D. 45,500

8) Which of the following shows the numbers in descending order?

$$\frac{1}{5}, 0.83, 68\%, \frac{5}{14}$$

A. $68\%, 0.83, \frac{1}{5}, \frac{5}{14}$

C. $0.83, 68\%, \frac{1}{5}, \frac{5}{14}$

B. $68\%, 0.83, \frac{5}{14}, \frac{1}{5}$

D. $\frac{1}{5}, \frac{5}{14}, 68\%, 0.83$

9) If 160% of a number is 128, then what is the 45% of that number?

A. 36

C. 63

B. 22.5

D. 25.5

10) Which of the following decimal numbers is less than $\frac{36}{32}$?

A. 1.152

C. 1.125

B. 1.065

D. 1.250

11) Mr. Jones saves $1,800 out of his monthly family income of $46,800. What fractional part of his income does he save?

A. $\frac{1}{26}$

C. $\frac{5}{24}$

B. $\frac{1}{13}$

D. $\frac{3}{26}$

12) 16% of what number is equal to 192?

    A. 30.72   C1,400

    B. 161.28   D. 1,200

13) 45 students took an exam and 36 of them failed. What percent of the students passed the exam?

    A. 35%   C. 65%

    B. 80%   D. 20%

14) A bank is offering 2.8% simple interest on a savings account. If you deposit $22,000, how much interest will you earn in seven years?

    A. $3,696   C. $4,200

    B. $4,312   D. $4,230

$$\frac{9}{20}, 0.453\%, 0.0453$$

15) Which of the following correctly orders the values above from least to greatest?

    A. $0.453\%, \frac{9}{20}, 0.0453$ C. $\frac{9}{20}, 0.453\%, 0.0453$

    B. $0.0453, 0.453\% \frac{9}{20}$ D. $\frac{9}{20}, 0.0453, 0.453\%$

16) Which of the following is equivalent to $\frac{2}{5}$?

    A. 0.04   C. 0.40

    B. 0.45   D. 4.0

17) If a gas tank can hold 32 gallons, how many gallons does it contain when it is $\frac{3}{4}$ full?

A. 26   C. 22.5

B. 18   D. 24

18) What is 0.2143 rounded to the nearest hundredth?

A. 0.214   C. 0.21

B. 0.215   D. 0.22

19) Jack earns $840 for his first 35 hours of work in a week and is then paid 1.5 times his regular hourly rate for any additional hours. This week, Jack needs $2,100 to pay his rent, bills and other expenses. How many hours must he work to make enough money in this week?

A. 24   C. 56

B. 35   D. 59

20) $2\frac{3}{4} - \frac{5}{8} = ?$

A. 1.152   C. 0.2125

B. 2.125   D. 2.152

STOP

This is the End of Section 1 of Test 2. You may check your work on this Section if you still have time.

Next-Generation Accuplacer Practice Test 2

Section 2

Quantitative Reasoning, Algebra, and Statistics

❖ **20 Questions.**

❖ **Total time for this test: No time limit**.

❖ **You may NOT use a calculator on this Section.**

(On a real Accuplacer test, there is an onscreen calculator to use on some questions).

Administered *Month Year*

1) Simplify $10x^4y^4(2x^2y^3)^3 =$

    A. $40x^9y^{10}$ C. $80\ x^{10}y^{10}$

    B. $40x^9y^{13}$ D. $80\ x^{10}y^{13}$

2) The perimeter of a rectangular yard is 300 meters. What is its length if its width is five time its length?

    A. 25 meters   C. 30 meters

    B. 15 meters   D. 60 meters

3) What is the equivalent temperature of 59°F in Celsius?

$$C = \frac{5}{9}(F - 32)$$

    A. 18   C. 20

    B. 15   D. 22

4) Right triangle ABC has two legs of lengths 10cm (AB) and 24cm (AC). What is the length of the third side (BC)?

    A. 18 cm   C. 37 cm

    B. 28 cm   D. 26 cm

5) Which of the following points lies on the line $2x + 3y = 7$?

    A. $(2, 1)$ C. $(-2, 1)$

    B. $(5, 1)$ D. $(3, -2)$

6) What is the area of a square whose diagonal is 8?

   A. 64   C. 42

   B. 32   D. 16

7) Two dice are thrown simultaneously, what is the probability of getting a sum of 4 or 6?

   A. $\frac{7}{36}$ C. $\frac{5}{9}$

   B. $\frac{2}{9}$ D. $\frac{3}{18}$

8) The mean of 40 test scores was calculated as 55. But it turned out that one of the scores was misread as 77 but it was 37. What is the correct mean of the test scores?

   A. 52   C. 54

   B. 45   D. 56

9) In a stadium the ratio of home fans to visiting fans in a crowd is 4:9. Which of the following could be the total number of fans in the stadium?

   A. 11,219   C. 24,729

   B. 13,453   D. 56,840

10) The perimeter of the trapezoid below is 60 cm. What is its area?

   A. 214 cm^2

   B. 204 cm^2

   C. 221 cm^2

   D. 187 cm^2

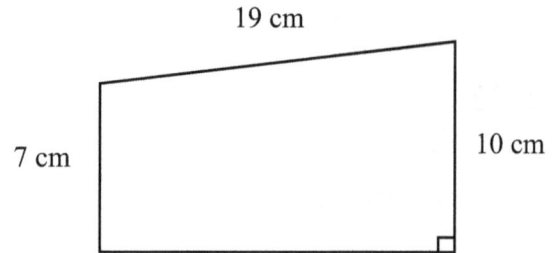

19 cm

7 cm

10 cm

11) A swimming pool holds 1,800 cubic feet of water. The swimming pool is 25 feet long and 6 feet wide. How deep is the swimming pool?

   A. 8   C. 14

   B. 15   D. 12

12) Anita's trick–or–treat bag contains 9 pieces of chocolate, 14 suckers, 26 pieces of gum, 15 pieces of licorice. If she randomly pulls a piece of candy from her bag, what is the probability of her pulling out a piece of sucker?

   A. $\frac{7}{32}$ C. $\frac{5}{32}$

   B. $\frac{14}{65}$ D. $\frac{15}{64}$

13) The average of 9 numbers is 20. The average of 6 of those numbers is 17. What is the average of the other three numbers?

   A. 39   C. 22

   B. 32   D. 26

14) A card is drawn at random from a standard 52–card deck, what is the probability that the card is of hearts and diamonds? (The deck includes 13 of each suit clubs, diamonds, hearts, and spades)

  A. $\frac{1}{52}$ C. $\frac{1}{13}$

  B. $\frac{1}{2}$ D. $\frac{2}{13}$

15) A football team had \$53,000 to spend on supplies. The team spent \$45,000 on new balls. New sport shoes cost \$280 each. Which of the following inequalities represent the number of new shoes the team can purchase?

  A. $280x + 45,000 \le 53,000$ C. $45,000x + 280 \le 53,000$

  B. $280x + 45,000 \ge 53,000$ D. $45,000x + 280 \ge 53,000$

16) The average of five numbers is 30. If a sixth number that is greater than 60 is added, then, which of the following could be the new average? (Select one or more answer choices)

  A. 30   C. 34

  B. 32   D. 36

17) The length of a rectangle is 3 meters less than 8 times its width. The perimeter of the rectangle is 102 meters. What is the area of the rectangle in meters?

  A. 39   C. 210

  B. 270   D. 240

18) The ratio of boys and girls in a class is 4:11. If there are 105 students in the class, how many more boys should be enrolled to make the ratio 1:1?

A. 28

B. 77

C. 49

D. 46

19) What is the value of x in the following equation?

$$\frac{4}{7}x + \frac{1}{6} = \frac{1}{2}$$

A. 7

B. $\frac{5}{6}$

C. $\frac{5}{12}$

D. $\frac{7}{12}$

20) What is the surface area of the cylinder below?

A. $280\ \pi\ \text{in}^2$

B. $160\ \pi\ \text{in}^2$

C. $480\ \pi\ \text{in}^2$

D. $240\ \pi\ \text{in}^2$

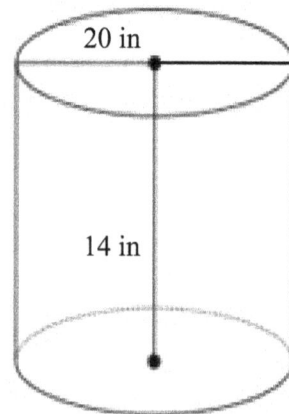

20 in

14 in

STOP

This is the End of Section 2 of Test 2. You may check your work on this Section if you still have time.

Next-Generation Accuplacer Practice Test 2

Section 3

Advanced Algebra and Functions

- ❖ **20 Questions.**

- ❖ **Total time for this test: No time limit**.

- ❖ **You may use a calculator on this Section.**

  (On a real Accuplacer test, there is an onscreen calculator to use on some questions).

Administered *Month Year*

1) Which of the following is equal to $b^{\frac{2}{9}}$?

A. $\sqrt{b^{\frac{2}{9}}}$

B. $\sqrt[2]{b^9}$

C. $\sqrt[9]{b^2}$

D. $\dfrac{1}{b^{\frac{2}{9}}}$

2) What is the value of $\dfrac{18b}{c}$ when $\dfrac{c}{b} = 6$

A. 18

B. $\dfrac{1}{3}$

C. 3

D. $\dfrac{1}{18}$

3) If $\dfrac{a-b}{b} = \dfrac{3}{7}$, then which of the following must be true?

A. $\dfrac{a}{b} = \dfrac{13}{17}$

B. $\dfrac{a}{b} = \dfrac{10}{7}$

C. $\dfrac{a}{b} = \dfrac{17}{7}$

D. $\dfrac{a}{b} = \dfrac{14}{7}$

4) Which of the following lines is parallel to $9y - 3x = 18$?

A. $y = \dfrac{1}{3}x + 10$

B. $y = 3x + 18$

C. $y = 3x - 5$

D. $y = 9x - 7$

5) What is the average of $4x + 3, -4x - 9$ and $9x + 4$?

A. $3x - \dfrac{2}{3}$

B. $3x + 2$

C. $2x - 3$

D. $3x - \dfrac{3}{4}$

6) If $\dfrac{x-3}{4} = N$ and $N = 5$, what is the value of x?

A. 17

B. 7

C. 23

D. 22

7) On Saturday, Sara read N pages of a book each hour for 4 hours, and Mary read M pages of a book each hour for 7 hours. Which of the following represents the total number of pages of book read by Sara and Mary on Saturday?

A. 28MN

C. 11MN

B. $4N + 7M$

D. $7N + 4M$

8) For $i = \sqrt{-1}$, which of the following is equivalent of $\frac{5+3i}{4-2i}$?

A. $0.7 + 1.5i$

C. $\frac{7+11i}{5}$

B. $\frac{7+11i}{10}$

D. $\frac{4+12i}{12}$

9) If function is defined as $f(x) = bx^2 + 11$, and b is a constant and $f(2) = 27$. What is the value of $f(-1)$?

A. 15

C. 17

B. 7

D. 5

10) Calculate $f(-2)$ for the function $f(x) = x^3 + 18$.

A. -10

C. 10

B. -26

D. 26

11) Find the solution (x, y) to the following system of equations?

$$x - 6y = 5$$
$$-2x + 3y = 8$$

A. $(1, -7)$

C. $(-7, 2)$

B. $(-7, -2)$

D. $(7, -2)$

12) What is the sum of all values of n that satisfies $2n^2 + 24n + 70 = 0$?

A. -2

C. 12

B. 2

D. -12

13) For $i = \sqrt{-1}$, what is the value of $\frac{5+i}{1+3i}$?

A. $\frac{4}{5}$

C. $\frac{7-4i}{5}$

B. $\frac{4-7i}{5}$

D. $4 - 7i$

$$y = x^2 - 12x + 32$$

14) The equation above represents a parabola in the xy-plane. Which of the following equivalent forms of the equation displays the x-intercepts of the parabola as constants or coefficients?

A. $y = (x - 4)(x - 8)$

C. $y = (x + 4)(x - 8)$

B. $y = (x + 4)(x + 8)$

D. $y = 4x(x - 8)$

15) The function $g(x)$ is defined by a polynomial. Some values of x and $g(x)$ are shown in the table below. Which of the following must be a factor of $g(x)$?

A. $x - 7$

B. $x - 6$

C. $2x - 6$

D. $x + 6$

x	$g(x)$
0	-6
2	-4
6	0

16) Which of the following is equivalent to $\frac{4x^2+(2x)^3+(x)^4}{x^2}$?

A. $8x^2 + 4x + 1$

C. $8x^2 + 4x$

B. $x^2 + 8x + 4$

D. $x^3 + 8x^2 + 4x$

17) A plant grows at a linear rate. After six weeks, the plant is 54 cm tall. Which of the following functions represents the relationship between the height (y) of the plant and number of weeks of growth (x)?

A. $y(x) = 54x + 9$

C. $y(x) = 54x$

B. $y(x) = 9x + 54$

D. $y(x) = 9x$

18) Sara orders a box of pen for $20 per box. A tax of 5.4% is added to the cost of the pens before a flat shipping fee of $6 closest out the transaction. Which of the following represents total cost of p boxes of pens in dollars?

A. $2.11p + 6$

C. $21.08p + 6$

B. $2.11p - 6$

D. $22.1p + 6$

19) A rectangle was altered by increasing its length by 22 percent and decreasing its width by s percent. If these alterations decreased the area of the rectangle by 8.5 percent, what is the value of s.

A. 30

C. 25

B. 10

D. 35

20) Tickets for a talent show cost $2 for children and $5 for adults. If John spends at least $23 but no more than $27 on x children tickets and 4 adult ticket, what are two possible values of x?

A. 4, 2

C. 3, 5

B. 2, 3

D. 3, 1

STOP

This is the End of this Test. You may check your work on this Test if you still have time.

Chapter 17 :

Answers and Explanations

Answer Key

✳ Now, it's time to review your results to see where you went wrong and what areas you need to improve!

Next- Generation Accuplacer Math Practice Test

Practice Test 1						Practice Test 2					
Section 1		**Section 2**		**Section 3**		**Section 1**		**Section 2**		**Section 3**	
1	C	1	B	1	B	1	A	1	D	1	C
2	B	2	B	2	C	2	D	2	A	2	C
3	A	3	C	3	C	3	C	3	B	3	B
4	D	4	B	4	A	4	D	4	D	4	A
5	C	5	A	5	C	5	A	5	A	5	A
6	C	6	C	6	C	6	C	6	B	6	C
7	A	7	C	7	C	7	C	7	B	7	B
8	C	8	C	8	A	8	D	8	C	8	B
9	D	9	B	9	D	9	A	9	A	9	A
10	B	10	D	10	A	10	B	10	B	10	C
11	D	11	A	11	B	11	A	11	D	11	B
12	C	12	D	12	B	12	D	12	A	12	D
13	C	13	D	13	C	13	D	13	D	13	B
14	B	14	D	14	C	14	B	14	B	14	A
15	C	15	D	15	A	15	D	15	A	15	B
16	D	16	C	16	B	16	C	16	D	16	B
17	C	17	B	17	D	17	D	17	B	17	D
18	C	18	B	18	C	18	C	18	C	18	C
19	C	19	D	19	D	19	D	19	D	19	C
20	B	20	A	20	C	20	B	20	C	20	B

Practice Tests 1
Section 1 - Arithmetic

1) Answer: C.

The closest to 8.12 is 8 in the options provided.

2) Answer: B.

$\frac{14}{5} = 2.8$, the only option that is greater than 2.8 is $\frac{9}{2} = 4.5$, $4.5 > 2.8$.

3) Answer: A.

First multiply the tenths place of 0.2 by 2.85. The result is 0.57. Next, multiply 8 by 2.85 which results in 22.8. The sum of these two numbers is: $0.57 + 22.8 = 23.37$.

4) Answer: D.

To compare fractions, find a common denominator. When two fractions have common denominators, the fraction with the larger numerator is the larger number. Choice A is incorrect because $\frac{16}{26}$ is not less than $\frac{7}{13}$. Write both fractions with common denominator and compare the numerators. $\frac{7}{13} = \frac{14}{26}$.

The fraction $\frac{16}{26}$ is greater than $\frac{7}{13}$.

Choice B and C are not correct. Shown written with a common denominator, the comparisons $\frac{5}{9} < \frac{1}{3}$ and $\frac{5}{6} < \frac{23}{42}$ are not correct. Shown written with a common denominator, the comparison $\frac{11}{24} < \frac{7}{8}$ is correct because $\frac{7}{8}$ or $\frac{21}{24}$ is greater than $\frac{11}{24}$.

5) Answer: C.

Dividing 54 by 45%, which is equivalent to 0.45, gives 120. Therefore, 45% of 120 is 54.

6) Answer: C.

Let x be the original price. If the price of the sofa is decreased by 38% to $279, then:

$62\%\ of\ x = 279 \Rightarrow 0.62x = 279 \Rightarrow x = 279 \div 0.62 = 450$.

7) Answer: A.

The percent of girls playing tennis is: $65\% \times 20\% = 0.65 \times 0.20 = 0.13 = 13\%$

8) Answer: C.

Write the equation and solve for B: $0.84\ A = 0.14\ B$, divide both sides by 0.14, then:

$\frac{0.84}{0.14}A = B$, therefore: $B = 6A$, and B is 6 times of A or it's 600% of A.

9) Answer: D.

$9 \times 10 = \$90$,　　Petrol use:$10 \times 1.5 = 15$ liters

Petrol cost: $15 \times \$3 = \$45$.　　Money earned: $\$90 - \$45 = \$45$

10) Answer: B.

Use this formula: Percent of Change $\dfrac{\text{New Value} - \text{Old Value}}{Old\ Value} \times 100\%$

$\dfrac{40,500 - 54,000}{54,000} \times 100\% = -25\%$ and $\dfrac{30,375 - 40,500}{40,500} \times 100\% = -25\%$

11) Answer: D.

First simplify the multiplication: $\dfrac{24}{7} \times \dfrac{14}{5} = \dfrac{48}{5} = \dfrac{48}{5}$

Choice D is equal to $\dfrac{48}{5}$.　　$\dfrac{24 \times 2}{5} = \dfrac{48}{5}$

12) Answer: C.

Use simple interest formula:

$I = prt$ ($I = interest,\ p = principal, r = rate, t = time$)

$I = (35,000)(0.025)(4) = 3,500$

13) Answer: C.

Let's compare each fraction: $\dfrac{4}{19} < \dfrac{5}{16} < \dfrac{7}{12} < \dfrac{8}{9}$　　Only choice C provides the right order.

14) Answer: B.

The question is this: 472.50 is what percent of 630? Use percent formula:

$part = \dfrac{\text{percent}}{100} \times whole$

$472.50 = \dfrac{\text{percent}}{100} \times 630 \Rightarrow 472.50 = \dfrac{\text{percent} \times 630}{100} \Rightarrow 47,250 = \text{percent} \times 630 \Rightarrow$

$\text{percent} = \dfrac{47,250}{630} = 75$

472.50 is 75% of 630. Therefore, the discount is:

$100\% - 75\% = 25\%$

15) Answer: C.

The weight of 21.4 meters of this rope is: $21.4 \times 500\ g = 10,700\ g$

1 kg = 1,000 g, therefore, $10,700\ g \div 1,000 = 10.7\ kg$

16) Answer: D.

$\dfrac{19}{6} - \dfrac{28}{6} = -\dfrac{9}{6} = -\dfrac{3}{2} = -1.5$

17) Answer: C.

61 divided by 7, the remainder is 5. 41 divided by 9, the remainder is also 5.

18) Answer: C.

Use percent formula: $\text{part} = \frac{\text{percent}}{100} \times \text{whole}$

$24 = \frac{\text{percent}}{100} \times 60 \Rightarrow 24 = \frac{\text{percent} \times 60}{100} \Rightarrow 24 = \frac{\text{percent} \times 6}{10}$, multiply both sides by 10.

$240 = \text{percent} \times 6$, divide both sides by 6. $40 = \text{percent}$

19) Answer: C.

Use distance formula: Distance = Rate × time $\Rightarrow 620 = 50 \times T$, divide both sides by 50. $\frac{620}{50} = T \Rightarrow T = 12.4$ hours.

Change hours to minutes for the decimal part. 0.4 hours = 0.4 × 60 = 24 minutes.

20) Answer: B.

To add decimal numbers, line them up and add from right.

$7.51 + 0.234 + 0.6367 = 8.3807$

Practice Tests 1

Section 2 - Quantitative Reasoning, Algebra, And Statistics

1) Answer: B.

Let x be the number. Write the equation and solve for x.

$\frac{4}{7} \times 35 = \frac{5}{9}.x \Rightarrow \frac{4 \times 35}{7} = \frac{5x}{9}$, use cross multiplication to solve for x.

$9 \times 4 \times 35 = 5x \times 7 \Rightarrow 1,260 = 35x \Rightarrow x = 36$

2) Answer: B.

Use the information provided in the question to draw the shape.

Use Pythagorean Theorem: $a^2 + b^2 = c^2$

$72^2 + 30^2 = c^2 \Rightarrow 5,184 + 900 = c^2 \Rightarrow 6,084 = c^2 \Rightarrow c = 78$

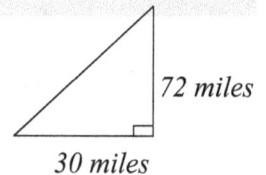
72 miles
30 miles

3) Answer: C.

The ratio of boy to girls is 3:7. Therefore, there are 3 boys out of 10 students. To find the answer, first divide the total number of students by 10, then multiply the result by 3.

$90 \div 10 = 9 \Rightarrow 9 \times 3 = 27$ There are 27 boys and 63 (90 – 27) girls. So, 36 more boys should be enrolled to make the ratio 1:1

4) Answer: B.

If the score of Mia was 80, therefore the score of Ava is 40. Since, the score of Emma was quarter as that of Ava, therefore, the score of Emma is 10.

5) Answer: A.

The sum of supplement angles is 180. Let x be that angle. Therefore, $x + 5x = 180$

$6x = 180$, divide both sides by 6: $x = 30$

6) Answer: C.

Let x be the number. Write the equation and solve for x. $(70 - x) \div x = 9$

Multiply both sides by x. $(70 - x) = 9x$, then add x both sides. $70 = 10x$, now divide both sides by 10. $x = 7$

7) Answer: C.

The average speed of john is: $150 \div 5 = 30 \ km$

The average speed of Alice is: $280 \div 4 = 70 \ km$

Write the ratio and simplify. $30:70 \Rightarrow 3:7$

8) Answer: C.

$5 \times 12 = \$60$,   Petrol use: $12 \times 2.5 = 30$ liters

Petrol cost: $30 \times \$1.2 = \$36$

Money earned: $\$60 - \$36 = \$24$

9) Answer: B.

The area of the trapezoid is:

Area$= \frac{1}{2}h(b_1 + b_2) = \left(\frac{11+16}{2}\right) \times x = 162 \rightarrow 13.5x = 162 \rightarrow x = 12$

$y = \sqrt{5^2 + 12^2} = \sqrt{169} = 13$

Perimeter is: $13 + 5 + 11 + 12 + 11 = 52$

10) Answer: D.

Use Pythagorean Theorem: $a^2 + b^2 = c^2$

$15^2 + 36^2 = c^2 \Rightarrow 225 + 1,296 = c^2 \Rightarrow 1,521 = c^2 \Rightarrow c = 39$

11) Answer: A

Area of the circle is less than than 169π. Use the formula of areas of circles.

Area $= \pi r^2 \Rightarrow 169\pi > \pi r^2 \Rightarrow 169 > r^2 \Rightarrow r < 13$

Radius of the circle is less than 13. Let's put 12 for the radius. Now, use the circumference formula:

Circumference $= 2\pi r = 2\pi (13) = 26\pi$

Since the radius of the circle is less than 13. Then, the circumference of the circle must be less than 26π. Only choice A is less than 26π.

12) Answer: D.

If the length of the box is 120, then the width of the box is one sixth of it, 20, and the height of the box is 5 (one fourth of the width). The volume of the box is:

V = (length) × (width) × (height) $= lwh = (120) \times (20) \times (5) = 12,000 \ cm^3$

13) Answer: D.

To find the number of possible outfit combinations, multiply number of options for each factor: $6 \times 7 \times 5 = 210$

14) Answer: D.

The equation of a line is in the form of $y = mx + b$, where m is the slope of the line and b is the $y - intercept$ of the line.

Two points $(2, 2)$ and $(4, 6)$ are online A. Therefore, the slope of the line A is:

slope of line A $= \frac{y_2 - y_1}{x_2 - x_1} = \frac{6-2}{4-2} = \frac{4}{2} = 2$

The slope of line A is 2. Thus, the formula of the line A is:

$y = mx + b = 2x + b$, choose a point and plug in the values of x and y in the equation to solve for b. Let's choose point $(2, 2)$. Then:$y = 2x + b \rightarrow 2 = 4 + b \rightarrow b = 2 - 4 = -2$

The equation of line A is: $y = 2x - 2$

Now, let's review the choices provided:

A. $(-1, 5)$ $y = 2x - 2 \rightarrow 5 = -2 - 2 = -4$ This is not true.

B. $(2, 6)$ $y = 2x - 2 \rightarrow 6 = 4 - 2 = 2$ This is not true.

C. $(-2, -3)$ $y = 2x - 2 \rightarrow -3 = -4 - 2 = -6$ This is not true.

D. $(-5, -12)$ $y = 2x - 2 \rightarrow -12 = -10 - 2 = -12$ This is true!

15) Answer: D.

If 12 balls are removed from the bag at random, there will be one ball in the bag.

The probability of choosing a brown ball is 1 out of 15. Therefore, the probability of not choosing a brown ball is 8 out of 15 and the probability of having not a brown ball after removing 8 balls is the same.

16) Answer: C.

Let x be the smallest number. Then, these are the numbers: $x, x + 1, x + 2, x + 3, x + 4$,

average $= \frac{\text{sum of terms}}{\text{number of terms}} \Rightarrow 84 = \frac{x+(x+1)+(x+2)+(x+3)+(x+4)}{5} \Rightarrow 84 = \frac{5x+10}{5} \Rightarrow 420 = 5x + 10$

$\Rightarrow 410 = 5x \Rightarrow x = 82$

17) Answer: B.

average $= \dfrac{\text{sum of terms}}{\text{number of terms}}$

The sum of the weight of all girls is: $18 \times 45 = 810 \ kg$

The sum of the weight of all boys is: $32 \times 55 = 1,760 \ kg$

The sum of the weight of all students is: $810 + 1,760 = 2,570 \ kg$

average $= \frac{2,570}{50} = 51.40$

18) Answer: B.

Write the numbers in order: 10, 12, 18, 23, 24, 29, 37

Since we have 7 numbers (7 is odd), then the median is the number in the middle, which is 23.

19) Answer: D.

Formula for the Surface area of a cylinder is:

$$SA = 2\pi r^2 + 2\pi rh \rightarrow 160\pi = 2\pi r^2 + 2\pi r(2) \rightarrow r^2 + 2r - 80 = 0$$

$$(r + 10)(r - 8) = 0 \rightarrow r = 8 \quad or \quad r = -10 \ (unacceptable)$$

20) Answer: A.

Let x be the number of years. Therefore, \$5,000 per year equals $5,000x$. Starting from \$36,000 annual salary means you should add that amount to $5,000x$.

Income more than that is: $I > 5,000x + 36,000$

Practice Tests 1

Section 3 - Advanced Algebra and Functions

1) Answer: B.

The input value is -2. Then: $x = -2$

$f(x) = 3x^2 - 5x + 2 \rightarrow f(-2) = 3(-2)^2 - 5(-2) + 2 = 12 + 10 + 2 = 24$

2) Answer: C.

Multiplying each side of $\frac{15}{x} = \frac{20}{x-5}$ by $x(x-5)$ gives $15(x-5) = 20(x) \Longrightarrow 3(x-5) = 4(x)$,

distributing the 2 over the values within the parentheses yields $3x - 15 = 4x$ or $x = -15$.

Therefore, the value of $\frac{x}{5} = \frac{-15}{5} = -3$.

3) Answer: C.

Method 1: Plugin the values of x and y provided in the options into both equations.

A. $(2, 3)$ $x - y = 0 \rightarrow 2 - 3 \neq 0$

B. $(-4, 1)$ $x - y = 0 \rightarrow -4 - 1 \neq 0$

C. $(3, 3)$ $x - y = 0 \rightarrow 3 - (3) = 0$

D. $(-1, 5)$ $x - y = 0 \rightarrow -1 - (5) \neq 0$

Only option C is correct.

Method 2: Multiplying each side of $x - y = 0$ by 5 gives $5x - 5y = 0$. Then, adding the corresponding side of $5x - 5y = 0$ and $2x + 5y = 21$ gives $7x = 21$. Dividing each side of $7x = 21$ by 7 gives $x = 3$. Finally, substituting 3 for x in $x - y = 0$, or $y = 3$. Therefore, the solution to the given system of equations is $(3, 3)$.

4) Answer: A.

If $f(x) = 4x + 3(x + 2) - 5$, then find $f(3x)$ by substituting $3x$ for every x in the function. This gives: $f(3x) = 4(3x) + 3(3x + 2) - 5$

It simplifies to: $f(3x) = 4(3x) + 3(3x + 2) - 5 = 12x + 9x + 6 - 5 = 21x + 1$

5) Answer: C.

First, find the equation of the line. All lines through the origin are of the form $y = mx$, so the equation is $y = \frac{1}{5}x$. Of the given choices, only choice C (10, 2), satisfies this equation:

$y = \frac{1}{5}x \rightarrow 2 = \frac{1}{5}(10) = 2$.

6) Answer: C.

$(7n^2 + 9n + 13) - (6n^2 + 5)$ Add like terms together: $7n^2 - 6n^2 = n^2$

$9n$ doesn't have like terms. $13 - (5) = 8$

Combine these terms into one expression to find the answer: $n^2 + 9n + 8$

7) Answer: C.

You can find the possible values of a and b in $(ax + 5)(bx + 4)$ by using the given equation $a + b = 10$ and finding another equation that relates the variables a and b. Since $(ax + 5)(bx + 4) = 21x^2 + cx + 20$, expand the left side of the equation to obtain

$abx^2 + 5bx + 4ax + 20 = 21x^2 + cx + 20$

Since ab is the coefficient of x^2 on the left side of the equation and 21 is the coefficient of x^2 on the right side of the equation, it must be true that $ab = 21$

The coefficient of x on the left side is $5b + 4a$ and the coefficient of x in the right side is c.

Then: $5b + 4a = c$, $a + b = 10$, then: $a = 10 - b$

Now, plug in the value of a in the equation $ab = 21$. Then:

$$ab = 21 \rightarrow (10 - b)b = 21 \rightarrow 10b - b^2 = 21$$

Add $-10b + b^2$ both sides. Then: $b^2 - 10b + 21 = 0$

Solve for b using the factoring method. $b^2 - 10b + 21 = 0$

$\rightarrow (b - 3)(b - 7) = 0$

Thus, either $b = 3$ and $a = 7$, or $b = 7$ and $a = 3$. If $b = 3$ and $a = 7$, then

$5b + 4a = c \rightarrow 5(3) + 4(7) = c \rightarrow c = 43$. If $b = 7$ and $a = 3$, then, $5b + 4a = c \rightarrow$

$5(7) + 4(3) = c \rightarrow c = 47$.   Therefore, the two possible values for c are 43 and 47.

8) Answer: A.

To rewrite $\dfrac{1}{\frac{1}{x-5} + \frac{1}{x+2}}$, first simplify $\dfrac{1}{x-5} + \dfrac{1}{x+2}$.

$$\frac{1}{x-5} + \frac{1}{x+2} = \frac{1(x+2)}{(x-5)(x+2)} + \frac{1(x-5)}{(x+2)(x-5)} = \frac{(x+2)+(x-5)}{(x+2)(x-5)}$$

Then: $\dfrac{1}{\frac{1}{x-5} + \frac{1}{x+2}} = \dfrac{1}{\frac{(x+2)+(x-5)}{(x+2)(x-5)}} = \dfrac{(x-5)(x+2)}{(x-5)+(x+2)}$. (Remember, $\dfrac{1}{\frac{1}{x}} = x$)

This result is equivalent to the expression in choice A.

9) Answer: D.

First find the slope of the line using the slope formula. $m = \frac{y_2 - y_1}{x_2 - x_1}$

Substituting in the known information.

$(x_1, y_1) = (1, 5), \quad (x_2, y_2) = (-1, 7); \quad m = \frac{7-5}{-1-1} = \frac{2}{-2} = -1$

Now the slope to find the equation of the line passing through these points. $y = mx + b$

Choose one of the points and plug in the values of x and y in the equation to solve for b.

Let's choose point $(1, 5)$. Then: $y = mx + b \rightarrow 5 = -1(1) + b \rightarrow 5 = -1 + b \rightarrow b = 6$

The equation of the line is: $y = -x + 6$

Now, plug in the points provided in the choices into the equation of the line.

A. $(4, 1) \quad y = -x + 6 \rightarrow 1 = -1(4) + 6 \rightarrow 1 = 2; \quad$ This is NOT true.

B. $(2, 3) \quad y = -x + 6 \rightarrow 3 = -1(2) + 6 \rightarrow 3 = 4; \quad$ This is NOT true.

C. $(9, -1) \; y = -x + 6 \rightarrow -1 = -1(9) + 6 \rightarrow -1 = -3; \quad$ This is NOT true.

D. $(3, 3) \quad y = -x + 6 \rightarrow 3 = -1(3) + 6 \rightarrow 3 = 3; \quad$ This is true!

Therefore, the only point from the choices that lies on the line is $(3, 3)$.

10) Answer: A.

It is given that $g(4) = 2$. Therefore, to find the value of $f(g(4))$, then $f(g(4)) = f(2) = 14$

11) Answer: B.

Since $(0, 0)$ is a solution to the system of inequalities, substituting 0 for x and 0 for y in the given system must result in two true inequalities. After this substitution, $y > b + x$ becomes $0 > b$, and $y < x + a$ becomes $0 < a$. Hence, b is negative and a is positive.

Therefore, $a > b$.

12) Answer: B.

Area of the triangle is: $\frac{1}{2} AD \times BC$ and AD is perpendicular to BC. Triangle ADC is a $30°, 60°, 90°$ right triangle. The relationship among all sides of right triangle $30°, 60°, 90°$ is provided in the following triangle: In this triangle, the opposite side of 30° angle is half of the hypotenuse. And the opposite side of 60° is opposite of $30° \times \sqrt{3}$

$AC = 8$, then $AD = 4\sqrt{3}$

Area of the triangle ABC is: $\frac{1}{2} AD \times BC = \frac{1}{2} 4\sqrt{3} \times 8 = 16\sqrt{3}$

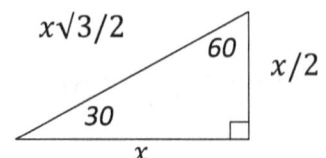

13) Answer: C.

It is given that $g(9) = 7$. Therefore, to find the value of $f(g(9))$, substitute 7 for $g(9)$.

$$f(g(9)) = f(7) = 26.$$

14) Answer: C.

The equation of a circle with center (h, k) and radius r is $(x - h)^2 + (y - k)^2 = r^2$. To put the

equation $x^2 + y^2 - 6x + 12y = 19$ in this form, complete the square as follows:

$x^2 + y^2 - 6x + 12y = 19 \Rightarrow (x^2 - 6x) + (y^2 + 12y) = 19$

$(x^2 - 6x + 9) - 9 + (y^2 + 12y + 36) - 36 = 19$

$(x - 3)^2 + (y + 6)^2 = 9 + 36 + 19 = 64$

$(x - 3)^2 + (y + 6)^2 = (8)^2$

Therefore, the radius of the circle is 8.

15) Answer: A.

By definition, the cosine of any acute angle is equal to the sine of its complement.

Since, angle A and B are complementary angles, therefore: $cos\ B = sin\ A$

16) Answer: B.

To solve this problem, first recall the equation of a line: $y = mx + b$

Where $m = slope.$ $y = y - intercept$

Remember that slope is the rate of change that occurs in a function and that the $y-$intercept is

the y value corresponding to $x = 0$.

Since the height of John's plant is 8 inches tall when he gets it. Time (or x) is zero. The plant

grows 4 inches per year. Therefore, the rate of change of the plant's height is 4. The $y-$intercept

represents the starting height of the plant which is 8 inches.

17) Answer: D.

The equation of a circle can be written as $(x - h)^2 + (y - k)^2 = r^2$

where (h, k) are the coordinates of the center of the circle and r is the radius of the circle.

Since the coordinates of the center of the circle are $(0, 4)$, the equation is $x^2 + (y - 4)^2 = r^2$,

where r is the radius. The radius of the circle is the distance from the center $(0, 4)$, to the

given endpoint of a radius, $\left(\frac{3}{2}, 2\right)$. By the distance formula, $r^2 = \left(\frac{3}{2} - 0\right)^2 + (2 - 4)^2 = \frac{25}{4}$

Therefore, an equation of the given circle is $x^2 + (y - 4)^2 = \frac{25}{4}$

18) Answer: C.

To solve for $\cos A$ first identify what is known.

The question states that $\triangle ABC$ is a right triangle whose $n\angle B = 90°$ and $\sin C = \frac{5}{13}$.

It is important to recall that any triangle has a sum of interior angles that equals 180 degrees.

Therefore, to calculate $\cos A$ use the complimentary angles identify of trigonometric function.

$\cos A = \cos(90 - C)$, Then: $\cos A = \sin C$

For complementary angles, sin of one angle is equal to cos of the other angle. $\cos A = \frac{5}{13}$

19) Answer: D.

In order to figure out what the equation of the graph is, fist find the vertex. From the graph we can determine that the vertex is at $(-3, 1)$.

We can use vertex form to solve for the equation of this graph.

Recall vertex form, $y = a(x - h)^2 + k$, where h is the x coordinate of the vertex, and k is the y coordinate of the vertex. Plugging in our values, you get $y = a(x + 3)^2 + 1$

To solve for a, we need to pick a point on the graph and plug it into the equation.

Let's pick $(0, 10)$, $10 = a(0 + 3)^2 + 1 \Longrightarrow 10 = a(3)^2 + 1 \Longrightarrow 10 = 9a + 1 \Longrightarrow 9a = 9$

$\Longrightarrow a = 1$;

Now the equation is: $y = 1(x + 3)^2 + 1$; Let's expand this, $y = (x^2 + 6x + 9) + 1$,

$y = x^2 + 6x + 10$. The equation in Choice D is the same.

20) Answer: C.

The line passes through the origin, $(5, m)$ and $(m, 15)$.

Any two of these points can be used to find the slope of the line. Since the line passes through $(0, 0)$ and $(5, m)$, the slope of the line is equal to $\frac{m-0}{5-0} = \frac{m}{5}$. Similarly, since the line passes through $(0, 0)$ and $(m, 15)$, the slope of the line is equal to $\frac{15-0}{m-0} = \frac{15}{m}$. Since each expression gives the slope of the same line, it must be true that $\frac{m}{5} = \frac{15}{m}$

Using cross multiplication gives

$\frac{m}{5} = \frac{15}{m} \rightarrow m^2 = 75 \rightarrow m = \pm\sqrt{75} = \pm\sqrt{25 \times 3} = \pm\sqrt{25} \times \sqrt{3} = \pm 5\sqrt{3}$

Practice Tests 2
Section 1 - Arithmetic

1) Answer: A.

To find the sum of two whole numbers, line the numbers up and add digits from right.
$$\begin{array}{r} 53,126 \\ +\ 7,492 \\ \hline 60,618 \end{array}$$

2) Answer: D.

15% of $420 is $0.15 \times 420 = 63$

3) Answer: C.

4 percent of $450 = \frac{4}{100} \times 450 = \frac{1}{25} \times 450 = \frac{1 \times 450}{25} = 18$

4) Answer: D.

The population is increased by 15% and 20%. 15% increase changes the population to 115% of original population. For the second increase, multiply the result by 120%.

$(1.15) \times (1.20) = 1.38 = 138\%$. 38 percent of the population is increased after two years.

5) Answer: A.

$90 off is the same as 30 percent off. Thus, 30 percent of a number is 90.

Then: $30\% \ of \ x = 90 \rightarrow 0.3x = 90 \rightarrow x = \frac{90}{0.3} = 300$

6) Answer: C.

The result when 530 is divided by 7 is 75 with a remainder of 5. Multiplying $7 \times 75 = 525$ and $530 - 525 = 5$, which is the remainder.

7) Answer: C.

Four times of 15,000 is 60,000. One eighth of them cancelled their tickets.

One eighth of 60,000 equal 7,500 ($\frac{1}{8} \times 60,000 = 7,500$).

$52,500(60,000 - 7,500 = 52,500)$ fans are attending this week.

8) Answer: D.

Change the numbers to decimal and then compare. $\frac{1}{5} = 0.2$;

0.83; $68\% = 0.68$; $\frac{5}{14} = 0.357$, Then: $\frac{1}{5} < \frac{5}{14} < 68\% < 0.83$

9) Answer: A.

First, find the number. Let x be the number. 160% of a number is 128, then: $1.6 \times x = 128 \Rightarrow x = 128 \div 1.6 = 80$; 45% of 80 is $0.45 \times 80 = 36$

10) Answer: B.

$\frac{36}{32} = 1.125$. The only choice less than 1.125 is 1.065.

$1.065 < 1.125 < 1.152 < 1.250$

11) Answer: A.

1,800 out of 46,800 equals to $\frac{1,800}{46,800} = \frac{18}{468} = \frac{1}{26}$

12) Answer: D.

Dividing 192 by 16%, which is equivalent to 0.16, gives 1,200.

13) Answer: D.

The failing rate is 36 out of 45 $= \frac{36}{45}$. Change the fraction to percent: $\frac{36}{45} \times 100\% = 80\%$

80 percent of students failed. Therefore, 20 percent of students passed the exam.

14) Answer: B.

Use simple interest formula: $I = prt$ (I = interest, p = principal, r = rate, t = time)

$I = (22,000)(0.028)(7) = 4,312$.

15) Answer: D.

To best compare the numbers, they should be put in the same format. The percent 0.453% can be converted to a decimal by dividing 0.453 by 100, which gives 0.00453. $\frac{9}{20}$ can be converted to a decimal by dividing 9 by 20, which gives 0.45. Now, all three numbers are in decimal format. $0.450 > 0.0453 > 0.00453$ or $\frac{9}{20} > 0.0453 > 0.453\%$, which is choice D.

16) Answer: C.

The fraction $\frac{2}{5}$ can be written as $\frac{2 \times 20}{5 \times 20} = \frac{40}{100}$, which can be interpreted as forty hundredths, or 0.40.

17) Answer: D.

$\frac{3}{4} \times 32 = \frac{96}{4} = 24$.

18) Answer: C.

The second digit to the right of the decimal point is in the hundredths place and the third number to the right of the decimal point is in the thousandths place. Since the number in the thousandths place of 0.2143, which is 4, is less than 5, the number 0.2143 should be rounded down to 0.21.

19) Answer: D.

The amount of money that jack earns for one hour: $\frac{\$840}{35} = \$24$

Number of additional hours that he works to make enough money is: $\frac{\$2,100-\$840}{1.5 \times \$24} = 35$

Number of total hours is: $24 + 35 = 59$

20) Answer: B.

$2\frac{3}{4} - \frac{5}{8} = \frac{11}{4} - \frac{5}{8} = \frac{22-5}{8} = \frac{17}{8} = 2.125$

Practice Tests 2

Section 2 - Quantitative Reasoning, Algebra, And Statistics

1) Answer: D.

Simplify. $10x^4y^4(2x^2y^3)^3 = 10x^4y^4(8x^6y^9) = 80x^{10}y^{13}$

2) Answer: A.

The width of the rectangle is twice its length. Let x be the length. Then, $width = 5x$

Perimeter of the rectangle is 2 (width + length) = $2(5x + x) = 300 \Rightarrow 12x = 300 \Rightarrow x = 25$

Length of the rectangle is 25 meters.

3) Answer: B.

Plug in 59 for F and then solve for C.

$C = \frac{5}{9}(F - 32) \Rightarrow C = \frac{5}{9}(59 - 32) \Rightarrow C = \frac{5}{9}(27) = 15$

4) Answer: D.

Use Pythagorean Theorem: $a^2 + b^2 = c^2$

$10^2 + 24^2 = c^2 \Rightarrow 676 = c^2 \Rightarrow c = 26$

5) Answer: A.

$2x + 3y = 7$. Plug in the values of x and y from choices provided. Then:

A. $(2, 1)$ $2x + 3y = 7 \rightarrow 2(2) + 3(1) = 7 \rightarrow 4 + 3 = 7$ This is true!

B. $(5, 1)$ $2x + 3y = 7 \rightarrow 2(5) + 3(1) = 7 \rightarrow 10 + 3 = 13$ This is NOT true!

C. $(-2, 1)$ $2x + 3y = 7 \rightarrow 2(-2) + 3(1) = 7 \rightarrow -4 + 3 = -1$ This is NOT true!

D. $(3, -2)$ $2x + 3y = 7 \rightarrow 2(3) + 3(-2) = 7 \rightarrow 6 - 6 = 0$ This is NOT true!

6) Answer: B.

The diagonal of the square is 8. Let x be the side.

Use Pythagorean Theorem: $a^2 + b^2 = c^2$

$x^2 + x^2 = 8^2 \Rightarrow 2x^2 = 8^2 \Rightarrow 2x^2 = 64 \Rightarrow x^2 = 32 \Rightarrow x = \sqrt{32}$

The area of the square is: $\sqrt{32} \times \sqrt{32} = 32$

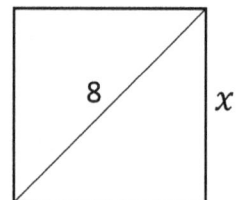

7) Answer: B.

To get a sum of 6 for two dice, we can get 5 different options:

$(1, 5), (5, 1), (2, 4), (4, 2), (3, 3)$

To get a sum of 4 for two dice, we can get 3 different options:

$(1, 3), (3, 1), (2, 2)$

Therefore, there are 8 options to get the sum of 4 or 6.

Since, we have $6 \times 6 = 36$ total options, the probability of getting a sum of 6 and 4 is 8 out of 36 or $\frac{8}{36} = \frac{2}{9}$.

8) Answer: C.

average (mean) $= \frac{\text{sum of terms}}{\text{number of terms}} \Rightarrow 55 = \frac{\text{sum of terms}}{40} \Rightarrow sum = 55 \times 40 = 2,200$

The difference of 77 and 37 is 40. Therefore, 40 should be subtracted from the sum.

$2,200 - 40 = 2,160$

mean $= \frac{\text{sum of terms}}{\text{number of terms}} \Rightarrow$ mean $= \frac{2,160}{40} = 54$

9) Answer: A.

In the stadium the ratio of home fans to visiting fans in a crowd is 4:9. Therefore, total number of fans must be divisible by 13: $4 + 9 = 13$.

Let's review the choices:

A. 11,219 $11,219 \div 13 = 863$

B. 13,453 $13,453 \div 13 = 1,034.85$

C. 24,729 $24,729 \div 13 = 1,902.23$

D. 56,840 $56,840 \div 13 = 4,372.31$

Only choice A when divided by 13 results a whole number.

10) Answer: B.

The perimeter of the trapezoid is 60 cm.

Therefore, the missing side (height) is $= 60 - 7 - 19 - 10 = 24$

Area of a trapezoid: $A = \frac{1}{2} h (b_1 + b_2) = \frac{1}{2}(24)(7 + 10) = 204$

11) Answer: D.

Use formula of rectangle prism volume.

V = (length)(width)(height) $\Rightarrow 1,800 = (25)(6)(\text{height}) \Rightarrow$

height $= 1,800 \div 150 = 12$

12) Answer: A.

Probability $= \frac{\textit{number of desired outcomes}}{\textit{number of total outcomes}} = \frac{14}{9+14+26+15} = \frac{14}{64} = \frac{7}{32}$

13) Answer: D.

$\text{average} = \frac{\text{sum of terms}}{\text{number of terms}} \Rightarrow$ (average of 9 numbers) $20 = \frac{\text{sum of numbers}}{9} \Rightarrow$ sum of 9 numbers is: $20 \times 9 = 180$

(average of 6 numbers) $17 = \frac{\text{sum of numbers}}{6} \Rightarrow$ sum of 6 numbers is: $17 \times 6 = 102$

sum of 9 numbers – sum of 6 numbers = sum of 3 numbers $180 - 102 = 78 \Rightarrow$ average of 3 numbers $= \frac{78}{3} = 26$

14) Answer: B.

The probability of choosing the diamond and hearts are $\frac{26}{52} = \frac{1}{2}$

15) Answer: A.

Let x be the number of new shoes the team can purchase. Therefore, the team can purchase $280x$.

The team had \$53,000 and spent \$45,000. Now the team can spend on new shoes \$8,000 at most.

Now, write the inequality: $280x + 45,000 \leq 53,000$

16) Answer: D.

First, find the sum of four numbers.

$\text{average} = \frac{\text{sum of terms}}{\text{number of terms}} \Rightarrow 30 = \frac{\text{sum of 5 numbers}}{5} \Rightarrow$ sum of 5 numbers $= 30 \times 5 = 150$

The sum of 5 numbers is 150. If a sixth number that is greater than 60 is added to these numbers, then the sum of 6 numbers must be greater than 210.

$150 + 60 = 210$. If the number was 60, then the average of the numbers is:

$\text{average} = \frac{\text{sum of terms}}{\text{number of terms}} = \frac{210}{6} = 35$

Since the number is bigger than 60. Then, the average of six numbers must be greater than 35. Choice D is greater than 35.

17) Answer: B.

Let L be the length of the rectangular and W be the width of the rectangular. Then,

$L = 8W - 3$

The perimeter of the rectangle is 102 meters. Therefore: $2L + 2W = 102$

$L + W = 51$

Replace the value of L from the first equation into the second equation and solve for W:

$(8W - 3) + W = 51 \rightarrow 9W - 3 = 51 \rightarrow 9W = 54 \rightarrow W = 6$

The width of the rectangle is 6 meters, and its length is:

$L = 8W - 3 = 8(6) - 3 = 45$

The area of the rectangle is: length × width = $45 \times 6 = 270$

18) Answer: C.

The ratio of boy to girls is 4:11. Therefore, there are 4 boys out of 15 students. To find the answer, first divide the total number of students by 15, then multiply the result by 4.

$105 \div 15 = 7 \Rightarrow 7 \times 4 = 28$

There are 28 boys and 77 (105 – 28) girls. So, 49 more boys should be enrolled to make the ratio 1:1

19) Answer: D.

Isolate and solve for x.

$\frac{4}{7}x + \frac{1}{6} = \frac{1}{2} \Rightarrow \frac{4}{7}x = \frac{1}{2} - \frac{1}{6} = \frac{1}{3} \Rightarrow \frac{4}{7}x = \frac{1}{3}$

Multiply both sides by the reciprocal of the coefficient of x. $\left(\frac{7}{4}\right)\frac{4}{7}x = \frac{1}{3}\left(\frac{7}{4}\right) \Rightarrow x = \frac{7}{12}$

20) Answer: C.

Surface Area of a cylinder = $2\pi r\,(r + h)$,

The radius of the cylinder is 10 (20 ÷ 2) inches and its height is 14 inches. Therefore,

Surface Area of a cylinder = $2\pi\,(10)\,(10 + 14) = 480\,\pi$

Practice Tests 2

Section 3 - Advanced Algebra and Functions

1) Answer: C.

$b^{\frac{m}{n}} = \sqrt[n]{b^m}$ For any positive integers m and n. Thus, $b^{\frac{2}{9}} = \sqrt[9]{b^2}$.

2) Answer: C.

To solve this problem first solve the equation for c. $\frac{c}{b} = 6$. Multiply by b on both sides. Then:

$b \times \frac{c}{b} = 6 \times b \to c = 6b$. Now to calculate $\frac{18b}{c}$, substitute the value for c into the denominator

and simplify. $\frac{18b}{c} = \frac{18b}{6b} = \frac{18}{6} = \frac{3}{1} = 3$

3) Answer: B.

The equation $\frac{a-b}{b} = \frac{3}{7}$ can be rewritten as $\frac{a}{b} - \frac{b}{b} = \frac{3}{7}$, from which it follows that $\frac{a}{b} - 1 = \frac{3}{7}$,

or $\frac{a}{b} = \frac{3}{7} + 1 = \frac{10}{7}$.

4) Answer: A.

First write the equation in slope intercept form. Add $3x$ to both sides to get $9y = 3x + 18$.

Now divide both sides by 9 to get $y = \frac{1}{3}x + 2$. The slope of this line is $\frac{1}{3}$, so any line that also

has a slope of $\frac{1}{3}$ would be parallel to it. Only choice A has a slope of $\frac{1}{3}$.

5) Answer: A.

To find the average of three numbers even if they're algebraic expressions, add them up and

divide by 3. Thus, the average equals: $\frac{(4x+3)+(-4x-9)+(9x+4)}{3} = \frac{9x-2}{3} = 3x - \frac{2}{3}$

6) Answer: C.

Since $N = 5$, substitute 5 for N in the equation $\frac{x-3}{4} = N$, which gives $\frac{x-3}{4} = 5$. Multiplying

both sides of $\frac{x-3}{4} = 5$ by 4 gives $x - 3 = 20$ and then adding 3 to both sides of

$x - 3 = 20$ then, $x = 23$.

7) Answer: B.

The total number of pages read by Sara is 4 (hours she spent reading) multiplied by her rate of

reading: $N \frac{pages}{hour} \times 4 hours = 4N$

Similarly, the total number of pages read by Mary is 7 (hours she spent reading) multiplied by her rate of reading: $M \frac{pages}{hour} \times 7hours = 7M$ the total number of pages read by Sara and Mary is the sum of the total number of pages read by Sara and the total number of pages read by Mary: $4N + 7M$.

8) Answer: B.

To rewrite $\frac{5+3i}{4-2i}$ in the standard form $a + bi$, multiply the numerator and denominator of $\frac{5+3i}{4-2i}$ by the conjugate, $4 + 2i$. This gives $\left(\frac{5+3i}{4-2i}\right)\left(\frac{4+2i}{4+2i}\right) = \frac{20+10i+12i+6i^2}{4^2-(2i)^2}$. Since $i^2 = -1$, this last fraction can be rewritten as $\frac{20+10i+12i+6(-1)}{16-4(-1)} = \frac{14+22i}{20} = \frac{7+11i}{10}$.

9) Answer: A.

First find the value of b, and then find $f(-1)$. Since $f(2) = 27$, substituting 2 for x and 27 for $f(x)$ gives $27 = b(2)^2 + 11 = 4b + 11$. Solving this equation gives $b = 4$. Thus $f(x) = 4x^2 + 11$,

$f(-1) = 4(-1)^2 + 11 \rightarrow f(-1) = 4 + 11, f(-1) = 15$

10) Answer: C.

Identify the input value. Since the function is in the form $f(x)$ and the question asks to calculate $f(-2)$, the input value is -2.

$f(-2) \rightarrow x = -2$, Using the function, input the desired x value.

Now substitute -2 in for every x in the function.

$f(x) = x^3 + 18, \quad f(-2) = (-2)^3 + 18, \quad f(-2) = -8 + 18, f(-2) = 10$

11) Answer: B.

Multiplying each side of $x - 6y = 5$ by 2 gives $2x - 12y = 10$. Adding each side of $2x - 12y = 10$ to the corresponding side of $-2x + 3y = 8$ gives $-9y = 18$ or $y = -2$. Finally, substituting -2 for y in $x - 6y = 5$ gives $x - 6(-2) = 5$ or $x = -7$.

12) Answer: D.

The problem asks for the sum of the roots of the quadratic equation $2n^2 + 24n + 70 = 0$. Dividing each side of the equation by 2 gives $n^2 + 12n + 35 = 0$. If the roots of $n^2 + 12n + 35 = 0$ are n_1 and n_2, then the equation can be factored as $n^2 + 12n + 35 = (n - n_1)(n - n_2) = 0$.

Looking at the coefficient of n on each side of $n^2 + 12n + 35 = (n + 5)(n + 7)$ gives $n = -5$ or $n = -7$, then, $-5 + (-7) = -12$

13) Answer: B.

To perform the division $\frac{5+i}{1+3i}$, multiply the numerator and

denominator of $\frac{5+i}{1+3i}$ by the conjugate of the denominator, $1 - 3i$. This gives $\frac{(5+i)(1-3i)}{(1+3i)(1-3i)} =$

$\frac{5-15i+i-3i^2}{1^2-9i^2}$. Since $i^2 = -1$, this can be simplified to $\frac{5-15i+i+3}{1+9} = \frac{8-14i}{10} = \frac{4-7i}{5}$

14) Answer: A.

The x-intercepts of the parabola represented by $y = x^2 - 12x + 32$ in the xy-plane are the

values of x for which y is equal to 0.

The factored form of the equation, $y = (x - 4)(x - 8)$, shows that y equals 0 if and only if

$x = 4$ or $x = 8$.

Thus, the factored form $y = (x - 4)(x - 8)$, displays the x-intercepts of the parabola as the

constants 4 and 8.

15) Answer: B.

If $x - a$ is a factor of $g(x)$, then $g(a)$ must equal 0. Based on the table $g(6) = 0$. Therefore,

$x - 6$ must be a factor of $g(x)$.

16) Answer: B.

Simplify the numerator. $\frac{4x^2+(2x)^3+(x)^4}{x^2} = \frac{4x^2+2^3x^3+x^4}{x^2} = \frac{4x^2+8x^3+x^4}{x^2}$. Pull an x^2 out of each

term in the numerator. $\frac{x^2(4+8x+x^2)}{x^2}$. The x^2 in the numerator and the x^2 in the denominator

cancel: $4 + 8x + x^2 = x^2 + 8x + 4$

17) Answer: D.

Rate of change (growth or x) is 9 per week. $54 \div 6 = 9$

Since the plant grows at a linear rate, then the relationship between the height (y) of the plant

and number of weeks of growth (x) can be written as: $y(x) = 9x$

18) Answer: C.

Since a box of pen costs $20, then $20p$ represents the cost of p boxes of pen.

Multiplying this number times 1.054 will increase the cost by the 5.4% for tax.

Then add the $6 shipping fee for the total: $1.054(20p) + 6$

19) Answer: C.

Let l and w be the length and width, respectively, of the original rectangle. The area of the original rectangle is $A = lw$. The rectangle is altered by increasing its length by 22 percent and decreasing its width by s percent; thus, the length of the altered rectangle is $1.22l$, and the width of the altered rectangle is $\left(1 - \frac{s}{100}\right)w$.

The alterations decrease the area by 8.5 percent, so the area of the altered rectangle is $(1 - 0.085)\,A = 0.915A$. The altered rectangle is the product of its length and width, therefore $0.915A = (1.22l)\left(1 - \frac{s}{100}\right)w$

Since $A = lw$, this equation can be rewritten as $0.915A = (1.22)\left(1 - \frac{s}{100}\right)lw = (1.22)\left(1 - \frac{s}{100}\right)$, from which it follows that $0.915 = (1.22)\left(1 - \frac{s}{100}\right)$, divide both sides of the equation by 1.22. Then: $0.75 = 1 - \frac{s}{100}$

Therefore, $\frac{s}{100} = 0.25$ and therefore the value of s is 25%.

20) Answer: B.

Because each child ticket costs \$2 and each adult ticket costs \$5, the total amount, in dollars, that John spends on x children tickets and 4 adult ticket is $2(x) + 4(5)$. Because John spends at least \$23 but no more than \$27 on the tickets, you can write the compound inequality $2x + 20 \geq 23$ and $2x + 20 \leq 27$.

Subtracting 20 from each side of both inequalities and then dividing each side of both inequalities by 2 gives $x \geq 1.5$ and $x \leq 3.5$. Thus, the value of x must be an integer that is both greater than or equal to 1.5 and less than or equal to 3.5. Therefore, $x = 2$ or $x = 3$. Either 2 or 3 may be gridded as the correct answer.

"End"

www.ingramcontent.com/pod-product-compliance
Lightning Source LLC
Chambersburg PA
CBHW081325090426
42737CB00017B/3033